T0251414

Water Policy in Australia

The Impact of Change and Uncertainty

Edited by

Lin Crase

New York   London

An RFF Press book
Published by Resources for the Future
2 Park Square, Milton Park, Abingdon, Oxon OX14 4RN
711 Third Avenue, New York, NY 10017

Library of Congress Cataloging-in-Publication Data

   Water policy in Australia : the impact of change and uncertainty / edited by Lin Crase. — 1st ed.
     p. cm.
   Includes bibliographical references and index.
   ISBN 978-1-933115-58-0 (hardcover : alk. paper)
   1. Water-supply—Government policy—Australia.   2. Water resources development—Government policy—Australia.   I. Crase, Lin.
   TD321.A1W39 2007
   363.6'10994—dc22   2007039653

The paper in this book meets the guidelines for permanence and durability of the Committee on Production Guidelines for Book Longevity of the Council on Library Resources. This book was typeset and copyedited by Integrated Book Technology. The cover was designed by Maggie Powell.

Cover photos, left to right: (1) Midlands Highway, Tasmania; (2) Kettering, Tasmania; (3) rolling irrigator in Murrumbidgee, New South Wales. Photos provided by Adam Gray and Jennifer McKay.

ISBN 978-1-933115-58-0

About Resources for the Future *and* RFF Press

Resources for the Future (RFF) improves environmental and natural resource policy-making worldwide through independent social science research of the highest caliber. Founded in 1952, RFF pioneered the application of economics as a tool for developing more effective policy about the use and conservation of natural resources. Its scholars continue to employ social science methods to analyze critical issues concerning pollution control, energy policy, land and water use, hazardous waste, climate change, biodiversity, and the environmental challenges of developing countries.

RFF Press supports the mission of RFF by publishing book-length works that present a broad range of approaches to the study of natural resources and the environment. Its authors and editors include RFF staff, researchers from the larger academic and policy communities, and journalists. Audiences for publications by RFF Press include all of the participants in the policymaking process—scholars, the media, advocacy groups, NGOs, professionals in business and government, and the public.

Resources for the Future

ISSUES IN WATER RESOURCE POLICY

Ariel Dinar, World Bank, Series Editor

Books in the *Issues in Water Resource Policy* series are intended to be accessible to a broad range of scholars, practitioners, policymakers, and general readers. Each book focuses on critical issues in water policy at a specific subnational, national, or regional level, with the mission to draw upon and integrate the best scholarly and professional expertise concerning the physical, ecological, economic, institutional, political, legal, and social dimensions of water use. The interdisciplinary approach of the series, along with an emphasis on real world situations and on problems and challenges that recur globally, are intended to enhance our ability to apply the full body of knowledge that we have about water resources—at local, country, and international levels.

For comments and editorial inquiries about the series, please contact *waterpolicy@rff.org*.
For information about other titles in the series, please visit *www.Rffpress.org/water*.
Cover photos, left to right: (1) Midlands Highway, Tasmania; (2) Kettering, Tasmania; (3) rolling irrigator in Murrumbidgee, New South Wales. Photos provided by Adam Gray and Jennifer McKay.

The *Issues in Water Resource Policy* series is dedicated to the memory of Guy LeMoigne, a founding member of the Advisory Committee.

To Phil Pagan (1969–2007)
Quiet scholar, committed colleague, good friend, and gentle soul.

Contents

Contributors

LIN CRASE is executive director of the Albury-Wodonga campus and associate professor of economics at La Trobe University, Victoria. He has spent over a decade researching the implications of different water property rights and commenting on the economic implications of water policy in Australia. His articles on these topics have appeared in the *Australian Journal of Agricultural and Resource Economics*, *Water Resources Research*, *Agenda*, and *The Australian Journal of Political Science*.

BRIAN DOLLERY is a professor of economics and director of the Centre for Local Government at the University of New England, Armidale. His research interests include public economics, microeconomics, and the economics of public policy. He has published numerous journal articles, books, and monographs in these areas.

RONLYN DUNCAN is a lecturer with the School of Geography and Environmental Studies at the University of Tasmania. Her research interests encompass environmental politics, environmental management and regulation, and the sociology of science. Before joining the university she worked for the New South Wales government's Sustainable Energy Development Authority.

GEOFF EDWARDS is an honorary associate in the School of Business at La Trobe University in Melbourne and is a visiting researcher at the Australian Productivity Commission. He is a former editor of the *Australian Journal of Agricultural and Resource Economics* and has published numerous articles on the economics of agricultural reform and the economics of regulation in *Agenda*, *The Economic Record*, and other scholarly outlets.

TERRY HILLMAN is adjunct professor of environmental management and ecology at La Trobe University. He was the director of the Murray-Darling Freshwater Research Centre and is a key advisor to state and national government agencies on the ecological implications of water policy changes.

AYNSLEY KELLOW is professor and head of the School of Government at the University of Tasmania. His research interests cover public policy, environmental policy, the politics of risk, electricity planning, trade and environment, and environmental administration.

STUART KHAN is a research fellow with the Centre for Water and Waste Technology at the University of New South Wales, Sydney. He is a leading Australian researcher on the trace organic chemical contaminants in water. Prior to joining UNSW, he held post-doctoral research positions at the University of Wollongong and the University of North Carolina.

REBECCA LETCHER is a research fellow at the Integrated Catchment Assessment and Management Centre at the Australian National University. She has authored multiple articles on concepts and approaches from various disciplinary areas for considering management issues such as water allocation and salinity. She is also the secretary of the International Environmental Modelling and Software Society.

JENNIFER MCKAY is a professor of business law at the University of South Australia. Her research interests concern water law reform and, in particular, regulatory models for the management and allocation of water between competing uses and between competing jurisdictions. She is the also the foundation director of the Centre for Comparative Water Policies and Laws at the university. She is a board member of the International Water Resource Association, serves on the editorial board of Water International UK, and is the Asia Pacific coordinator of the International Water Law Association.

JENNY MCLEOD is the environmental and communication manager for Murray Irrigation Limited. She has many years of practical experience dealing with policy changes in irrigation. She has been involved in the development and articulation of Murray Irrigation's response to a range of water policy issues including the Murray-Darling Basin Ministerial Council Cap on diversions, environmental flows for the Murray and Snowy Rivers, development of the Murray Lower Darling Water Sharing Plan, and bulk water price reviews.

WARREN MUSGRAVE is a consulting economist and a distinguished fellow of the Australian Agricultural and Resource Economics Society. From 1995 to 2000 he was special advisor in natural resources to the New South Wales Premier and is an emeritus professor of the University of New England, Armidale.

BLAIR NANCARROW is director of the Australian Research Centre for Water in Society. Her research includes determining the triggers of behavioral change in communities to achieve environmental sustainability in both urban and rural areas, and developing methods to incorporate human values and quality of life into sustainability assessments.

SUE O'KEEFE is deputy head of the School of Business at La Trobe University. Her research interests focus on decisionmaking and the role of economic incentives.

PHIL PAGAN was a doctoral candidate at the Centre for Resource and Environmental Studies at the Australian National University and a policy economist with the New South Wales Department of Primary Industries. His research focused on the evaluation of institutional arrangements for the interstate trading of water entitlements. As a policy economist, he had extensive experience with project evaluations related to irrigation sector structural adjustment.

SUSAN POWELL is pursuing her doctorate at the Fenner School of Environment and Society, Australian National University, specializing in remote sensing and modelling of wetland response to environmental flows.

JOHN QUIGGIN is a federation fellow in economics and political science at the University of Queensland. He has published articles, books, and reports on environmental economics, risk analysis, production economics, and the theory of economic growth. He has also written on policy topics including unemployment policy, micro-economic reform, privatisation, competitive tendering, and the management of the Murray-Darling river system.

JOHN ROLFE is professor of regional development economics at Central Queensland University where he is head of the environmental economics program. His research interests encompass environmental economics and environmental evaluation, particularly choice modelling and benefit transfer techniques. He has published several books and journal articles.

GEOFFREY J. SYME is research director for Society, Economy and Policy at CSIRO (Australian Commonwealth Scientific and Research Organization) Land and Water. His research relates to public involvement and fairness theory in water resource management. He is a member of the government Sustainability Assessment Panel for the development of the Yarragadee aquifer in Western Australia.

GEORGE WARNE is the general manager of Murray Irrigation Limited. He has been a strong public advocate for irrigation farmers and holds numerous positions on state and national water policy bodies. He has also served as national deputy-chairman of the Australian Committee on Irrigation and Drainage (ANCID) and as an irrigation industry representative on the Murray-Lower Darling Community Reference Committee.

CHAPTER 1

An Introduction to Australian Water Policy

Lin Crase

"Nothing is more useful than water: but it will purchase scarce anything; scarce anything can be had in exchange for it." (Adam Smith 1723–1790)

ADAM SMITH'S CONUNDRUM, which contrasted the usefulness of water with its value, was composed in the temperate surrounds of eighteenth-century Scotland. At that time, and in that location, relatively little attention was paid to the challenge of allocating water resources between competing users. Smith was unlikely, however, to have been able to predict the expanded demands on water resources in urban, agricultural, and industrial contexts arising from a global population explosion of Malthusian proportions. His vision of limited intervention by the state in economic affairs might also now be criticized in the context of the current dilemmas confronting water policy-makers. After all, water is a complex commodity and its allocation is circum-scribed by an array of social, historical, and political influences that com-plicate the development, implementation, and enforcement of public policy. Following Smith's maxim of the invisible hand of self interest and assuming that government should simply curtail its influence, runs the risk of denying this complexity. It also ignores the convoluted relationship that humans have with this most vital resource.

Notwithstanding the contextual limitations of Smith's observations, the need to coordinate the demands of water users had been well established in more arid parts of the world for centuries. For example, numerous indigenous com-munities in Australia had long regarded water as a common property resource and recognized the need for unique rules to govern its distribution (Coop and Brunckhorst 1999). Similarly, irrigators in the Banecher Valley in Spain, who gathered to sign their articles of association in May, 1435, created formal institu-tional arrangements to regulate water use (Ostrom 1992). More recently, how-ever, the task of finding appropriate policy frameworks for dealing with new and emerging conflicts over water usage is proving particularly challenging in many countries.

Arguably, these challenges are no more apparent than in Australia, which is often, and perhaps erroneously, cited as the driest settled continent on Earth (A. Watson 2003). Despite this rhetoric, such overt acceptance of the relative scarcity of the nation's water resources is a comparatively recent phenomenon. Historically, Australia's water resources have been seen as a factor of production to be harnessed in both agricultural and industrial contexts and as a vehicle for stimulating regional economic development. Water resources and their allocation were thus intrinsically tied to social and strategic objectives associated with nonmetropolitan economic development, such as closer settlement and soldier settlement (Langford-Smith and Rutherford 1966). This view broadly informed water policy formulation in Australia until the 1980s (B. Watson 1990). Only then was limited consideration given to the true opportunity cost of water usage and the potentially deleterious effects of particular extractions. Designing acceptable policy responses to these emerging issues has attracted the attention of Australian bureaucrats, politicians, scholars, and a range of other professionals for at least the past two decades.

Since the 1980s the development hypothesis has been supplanted by a management approach more consistent with the notion of a mature water economy. This water economy is characterized by an inelastic supply of new water and the need for expensive rehabilitation of aging projects (Randall 1981). The water management regime, congruous with a mature water economy, also requires policymakers to broaden the scope of their objectives to include economic efficiency, sustainable development, and ecological sustainability (B. Watson 1990). This is reflected in the increasing complexity of the allocation of water resources among multiple and competing environmental, economic, and social objectives. In the Australian context, the result has been a myriad of legislative water reforms accompanied by significant challenges for achieving social and economic adjustment. Learning to cope with ongoing water shortage has now assumed profound political, social, and economic importance in Australia and is thus worthy of analysis in its own right.

Australia has a water policy legacy that arose from a philosophy focusing primarily on resource development and extraction. Many of the institutional arrangements that circumscribe water reflect this earlier ethos and the wider cultural, economic, and social milieu. Thus, the challenge of developing contemporary water policy takes on important institutional dimensions. Moreover, reflecting on the evolution of Australian water policy is likely to prove particularly instructive for others engaged in similar processes worldwide. In this context, *The Economist* (2003) has proclaimed Australia as "the country that takes top prize for sensible water management." Despite such praise, some elements of Australian water policy remain the subject of criticism (see, for instance, Crase et al. 2004). Thus, understanding both the successes and the deficiencies of water policy may offer valuable insights transposable to other contexts.

The Scope of Australian Water Policy

Compiling a manuscript that deals with contemporary policy issues provides unique challenges. This volume represents the culmination of research, review, and analysis over a two-year period commencing in 2005. It is worth noting that 2006 represented the driest year on record for much of southern Australia. By the summer of 2006, many of the storages used to provide security of surface water were severely depleted, adding a sense of urgency to the water reform agenda. The water shortage simultaneously raised community consciousness of the potential impacts of anthropogenic climate change. The political landscape that circumscribed water resources at this time was also fluid, with the Commonwealth government offering to expand its responsibilities to encompass the management of water resources in the Murray-Darling Basin in early 2007. As 2007 unfolds water scarcity in the Murray-Darling Basin is at unprecedented levels and the pressure on policymakers is immense. Notwithstanding these emerging issues, the observations offered about water policy in this manuscript extend beyond this period and are grounded in more pervasive trends.

The extent to which these observations about Australian water reform can be generalized to other locales and different policy arenas is partly constrained by the inimitability of the Australian setting; however, its distinguishing characteristics also provide unique opportunities for comparative analyses between jurisdictions that vary widely in their physical and social makeup. In order to appreciate the extent of these contrasts, an (incomplete) sketch of the dynamics that circumscribe Australian water policy is offered here. This comprises an ambitious précis of the hydrological, physical, historical, political, social, and economic landscapes. In many cases, a more comprehensive treatment of these topics and their particular influence on water policy is provided in subsequent chapters of this book.

Physical Aspects and Indigenous Settlement

Australia is a land of stark physical contrasts. The island continent spans from the tropics in the north to cool temperate climates in the south. The land mass comprises almost 8 million square kilometers, representing the sixth-largest country in the world by area. Millennia of geographic isolation have resulted in unique natural fauna and flora, all well-adapted to withstand the vagaries of the Australian climate. The land mass itself is among the oldest and flattest, and around 20 percent is classified as desert whereas almost two-thirds is regarded as either semi-arid or arid and unsuitable for settlement by modern standards. Many of the ancient subsoils are also heavily invested with salt (Australian Academy of Science 2004). The largest store of salt exists in the sedimentary Murray Groundwater Basin, with groundwater stores often approaching the salinity of sea water (ABS 2002). However, the dominance of perennial vegetation in the natural landscape kept salinity in check for thousands of years.

Australia's indigenous population settled all parts of the landscape despite the harshness of climate. Estimates of the length of indigenous inhabitation vary with conservative calculations suggesting permanent aboriginal settlement for the last 40,000 years (Mulvaney and Kamminga 1999), although human remains discovered at Lake Mungo in southeast Australia have been dated at about 60,000 years (Thorne et al. 1999). Aboriginal culture is widely proclaimed as possessing a profound knowledge of water and a strong affinity with the environment.

Moreover, embedded social norms emphasized the rationale for a sustainable yield from resources. Magarey (1894–1895, *15*) acknowledged the extent of water management expertise held by indigenous Australians and the advantages of this knowledge bestowed on early European settlers as follows:

> "the white Australian may traverse the dry realms of arid Australia with all the confidence of his dusky brother, the Australian Aborigine who now is hovering on the horizon of an early extinction, leaving us as one permanent memento of his existence his legacy of lore in water quest."

Values held by the indigenous population were in stark contrast to those brought by European settlers, however. For instance, Lloyd (1988, 20) observed that "white settlers' [. . .] exploitation of water was often injudicious and inequitable [and they] were patronizing, even contemptuous, about Aboriginal water management." This exploitative approach radically altered the natural environment. Thus, while the history of European settlement in Australia is more concise than indigenous inhabitation, the impact on the natural landscape has been profound.

British Colonization and State Traditions

European settlement began in earnest in the late eighteenth century. Six British colonies were established over a forty-year period; initially commencing in the east of the mainland (New South Wales [NSW] was colonized in 1788), and concluding with the colonization of Western Australia in 1829. In contrast to the indigenous philosophy based on harmonization with the available resources, European settlement was typified by an attitude of "taming and transforming the land." In this context, however, interest in water for irrigation and industrial development did not seriously surface within the colonies until the late nineteenth and early twentieth centuries. Tasmania was the first colony to play an active part in irrigation development following the 1840–1843 droughts. Thereafter, other colonies became increasingly active in water resource development in response to episodic droughts and flooding accompanied by pressures from gold mining and pastoral development (Hallows and Thompson 1998, *17*).

One of the legacies of early autonomous colonial governance is the resulting state-based institutions that arose after federation at the beginning of the twentieth century. Many of these institutions continue to hold sway over national water policy formulation today.

First, each of the Australian states has retained superordinate legal status over water resources. Thus, while a hierarchical system of water rights might appropriately describe the Australian institutional status quo (Challen 2000), the influence of the states over the shape of water rights remains paramount. In this context colonial history has strongly prejudiced the conceptual and legal definition of water rights, with each state choosing its own interpretation of the appropriate replacement for British riparian traditions.

Secondly, differing ambitions for the use of water resources have resulted in significant enterprise differences between state jurisdictions. For instance, Victoria and NSW both originally sought to harvest water from the River Murray with a view to encouraging irrigation development. By way of contrast, South Australia had always sought to use the river as a means of transport, with the hope of establishing the Murray mouth as the major port from which to transport Murray-Darling Basin produce (Clark 1971).

These historically divergent prospects and approaches to water policy can still be distinguished today. Irrigation in NSW focuses on annual allocation of all available resources, and agriculture is accordingly dominated by annual cropping. By way of contrast, the Victorian approach to water allocation has always been more conservative, budgeting for longer-term water security. Ultimately, this has resulted in a preponderance of perennial irrigation enterprises, like horticulture and dairying. In South Australia, the original interest in navigation and its geographical location at the tail of the Murray-Darling catchment resulted in an even more conservative allocation regime. This was reinforced by the need to supplement Adelaide's water supply with extractions from the River Murray. The upshot has been the development of permanent horticultural enterprises making relatively frugal use of irrigation resources.

In comparison to other states with an interest in the Murray-Darling Basin, Queensland's water development history is relatively recent. Originally, attention centered on groundwater extraction, primarily as a means of securing stock and domestic supplies; however, by 1889, attention had turned to the development of groundwater to irrigate sugar cane (Hallows and Thompson 1998, *21*), and government-sponsored surface water schemes had emerged by 1922 with the establishment of the Dawson Valley scheme (DNRM 2002, *1*). In some respects the belated development of Queensland's water resources has resulted in it being out of step with the policies employed in other states. For instance, whereas the Victorian government has announced that there will be no new dams constructed within its jurisdiction (DSE 2004), enthusiasm for dam building remains a feature of the policy approach in Queensland. The Burnett River Dam in the north was completed in 2005 and two new dams are planned for construction in the southeastern part of the state.

Subsequent to early interest in irrigation in the 1840s, Tasmania's recent water development phase has been strongly influenced by that state's interest in hydroelectricity. During the 1960s the Tasmanian Hydroelectricity Commission developed plans to expand hydropower in order to attract energy-intensive industries to the state. Over the ensuing twenty years these plans were broadly implemented until controversial intervention by the federal government in 1983 halted the construction of the Franklin Dam. A rationale more akin to the maturing water economy has subsequently typified water resource management in this state.

Western Australia's water resource history is characterized by relatively modest interest in public irrigation, but for different reasons. Most early curiosity was focused on securing the limited water available in dry seasons for stock and domestic purposes. Irrigation developments in the southwest of the state commenced in the early twentieth century with the familiar goal of closer settlement and the added incentive to reduce Perth's reliance on food produce from the eastern states. Most other surface water irrigation developments are more recent and located in the north of the state (Hallows and Thompson 1998,

111). Groundwater resources play a more prominent role in Western Australian irrigation, supplying around 60 percent of demand (Johnson et al. 2005, 5). This has also driven the evolution of different institutional arrangements with a significant proportion of the state's horticulture developing from self-supplied groundwater and therefore relying less on state-sponsored investment than in other states. Much of the burden for urban and industrial demand is also met by groundwater in this state.

In summary, the legacy of ostensibly independent state control and governance of water resources should not be understated, and the resultant differences between states can be profound. Perhaps ironically, many earlier versions of state water legislation foresaw only modest direct intervention by the state and emphasized the need for private investment to foster infrastructure development (see, for instance, Lloyd 1988, 113–126). However, state-sponsored water infrastructure in some shape or form, particularly in irrigation, was firmly entrenched in all states by the end of the first decade of the twentieth century. Accompanied by differing views on the appropriate development of water resources, this has produced a complex array of state-based water policies.

Coordination Between States and Recent Policy Objectives

The strong foundation of the autonomy of states in the Australian constitution has made the task of producing coordinated action on water management problematic. Nevertheless, analysis of the structure and behavior of the institutions that have emerged to meet these challenges is instructive. Arguably, the area where coordination has received the most national attention is the Murray-Darling Basin, which spans several Australian states and territories, and a hydrological institutional model of catchment management has evolved to produce a coordinated response within this domain. Under this model, agreement between jurisdictions is required to permit a catchment-based approach to water management, particularly where the federal system is relatively weak (Green 2003, 130). The Murray-Darling Basin Ministerial Council (MDBMC) is the major governing body and comprises government ministers from NSW, Victoria, South Australia, Queensland, the Australian Capital Territory, and the Commonwealth. The operational arm of the MDBMC is the Murray-Darling Basin Commission (MDBC) and both organizations owe their existence to the Murray-Darling Basin Agreement, which is set out in the 1914 Murray-Darling Basin Act.

The Murray-Darling Basin extends from north of Roma in Queensland to Goolwa in South Australia. It also includes three-quarters of NSW, covers half of the land area of Victoria, and all of the Australian Capital Territory. In total, the Basin comprises over 1 million square kilometers, has a population of about 2 million, accounts for 40 percent of the national income derived from agriculture and almost three-quarters of the irrigated land within Australia (MDBC 1998). Partly because of its economic significance, the activities within the Basin have dominated recent water policy reforms in Australia. The Basin's close proximity to the major cities of Sydney, Melbourne, Adelaide, and Brisbane probably also accounts for its substantive influence over national water policy. Moreover, the purported national activities endorsed by the Council of Australian Governments (CoAG) reflect the emphasis on affairs within the Basin.

CoAG has played an increasingly influential role in setting national water policy over the last decade. Comprising representatives of all states, territories, and the Commonwealth government, CoAG was originally formed to progress competition reforms aimed at enhancing international competitiveness and improving efficiency. Importantly, it includes states that have little interest in the Murray-Darling Basin. The signing of the CoAG Agreement on Water Resource Policy (Water Reform Framework) in February 1994, and later the Competition Principles Agreement in April 1995, began the transformation of CoAG into one of the major authorities shaping Australian water policy. In essence, it now provides a vehicle for encouraging national, coordinated water management beyond the Murray-Darling Basin. At the core of CoAG's influence is the financial might of the Commonwealth government relative to the states. Progress against the agreed reform agenda is monitored and attracts financial payments from the Commonwealth to the state and territory governments. Failure to make adequate policy reform results in the withholding of transfer payments.

Despite this influence, to date, CoAG has not been able to completely supplant states' interests, and the institution itself remains a product of the political, economic, and historical landscape. For instance, CoAG announced in 2003 that it had agreed on a framework for developing a national water market. The enthusiasm for this reform was decidedly tepid in Western Australia and Tasmania and both states refused to sign the initiative (DPMC 2005). This is perhaps not surprising given that the water resources in these states are largely independent of others and interstate trade for these jurisdictions is nonsensical in the context of presently conceivable water prices.

In 2006 and 2007 the Commonwealth expanded its attempts to control the Murray-Darling Basin but geographic variations between states will remain, making the challenge of dealing with water policy at a national level formidable. Notwithstanding these political, social, historical, geographical, and hydrological realities, several enduring themes remain pertinent to Australian water policy. First, low rainfall and run-off, accompanied by significant variability, encouraged and continues to promote the perception among policymakers that Australia is a dry continent. Even the more humid states are perceived in this light. One of the consequences has been state-sponsored irrigation infrastructure which has often ignored the comparative advantage of dry land agriculture. As A. Watson (2003, 214) observed, " . . . the popular enthusiasm for irrigation over a century or more was always inconsistent with a serious appraisal of physical or economic opportunities facing Australia." This approach also overlooked the fact that Australia has more water per head of population than many other countries. Secondly, scant attention to economic fundamentals has manifested itself in subsidized production and the under-pricing of inputs, particularly water. Thirdly, the attempt to engineer settlement of the vast interior, including assigning returning soldiers to small unsustainable irrigated lots, occurred in all states. This was, at least partially, a response to the desire to water the land and transform semi-arid environs into something more akin to the European landscape. The resulting social infrastructure has made the task of redistribution of water resources on economic grounds even more problematic in this era.

Social and Environmental Parameters

At the social level it is important to recognize that the earlier public zeal for irrigation and water extraction is arguably an expression of the wider acceptance of the resource development ethos which typified the public policy settings in Australia over the past two centuries. That is, itself, a legacy of the early European traditions. Moreover, popular support for agrarian traditions persists within the Australian culture (see, for instance, Turner 2005), despite the fact that Australia is one of the most urbanized countries on earth. Political expressions of these values can be found in state-sponsored subsidization of rural industries, dedicated assistance programs for agricultural enterprises, publicly funded drought relief programs, preferential tax treatment for agricultural investments, and a range of specific programs targeted to aid rural and regional communities. Notwithstanding that this support is modest relative to that provided by other nations, establishing the appropriate price paid to the state for water used in agricultural and other uses remains problematic. Less overt expressions of support for agrarian traditions can be found elsewhere. For instance, images used to depict Australia to an international audience often borrow from rural life (Culture and Recreation Portal 2005). At the individual level many urban Australians don fashion items that have their roots in agricultural practices and the Australian lexicon continues to include (mostly compassionate) reference to *the bush*. Poignantly, this is defined as "any rural area, when contrasted with the city" (Macquarie Dictionary 2003, *133*). Similarly, Don Watson's humorous attempt to demystify the Australian language in his *Dictionary of Weasel Words* cites the bush as being "[t]he country; Rural Australia; Not the city; Not Sydney; That with which all Australians should be in touch" (D. Watson 2004, *52*). Notably, the distinction between those elements of the bush that remain primarily in their native form and those that have been radically altered as a result of agricultural pursuits is not at all clear in the vernacular. This may partially account for the ambiguity apparent in Australian water policy between irrigation efficiency and purported environmental enhancements.

Despite this vagueness, the values that Australians (both urban and rural) ascribe to the natural environment have unequivocally changed since the most energetic phases of the development era. There is now widespread acceptance that one of the legacies of our early enthusiasm to harness resources exclusively for productive or extractive purposes has been widespread environmental degradation (Crase et al. 2004). This is particularly the case with irrigation, where compelling evidence of the extent of environmental ills confronting the community is emerging. For instance, the MDBC (1998) has estimated that, on present trends, the salinity levels at Morgan on the River Murray will exceed the World Health Organization's desirable drinking water standard more than half of the time by 2020. Similarly, the MDBC (2002) has indicated that between 20 and 40 percent of the water currently extracted for irrigation from the River Murray will need to be returned if the river is to be restored to the status of a healthy working river. In addition, Western Australia is now characterized by large tracts of dry land salinity which have resulted from the expansion of agriculture into marginal lands (Pannell 2005). The security of Western Australia's capital city's water supply is also under threat with a realization that current extractions from groundwater aquifers are unsustainable (Kingsley 2002). In a related vein, environmental concerns continue to be expressed about the damming of waterways in Tasmania for hydroelectricity generation (see, for example, Blakers 1994).

Importantly, these challenges are not just environmental; they represent major social and economic adjustments for the Australian community. Centuries of ingrained acceptance of the merits of agricultural and other extractive pursuits are being challenged by our emerging knowledge of the precarious nature and peculiarity of the Australian landscape. Simultaneously, urban and industrial growth continues to outstrip the economic performance of agriculture. Where agriculture was once regarded as the foundation of national wealth, it now accounts for as little as 17 percent of export income and only 4.3 percent of Gross Domestic Product (ABS 2003). By way of contrast, in 1950 Australia was euphemistically 'riding on the sheep's back' with the boom in wool following the Korean War underpinning unprecedented prosperity and growth (ABS 2003). In the context of the emerging competition for water resources, the esteem ascribed to rural Australians by their urban cousins is itself being challenged. As water becomes increasingly scarce the case for continued support for extractive, relatively low-value production becomes weakened and the justification for the existence of rural communities is also under scrutiny (see, for instance, Kenyon and Black 2001).

In a broad sense, the Australian response to the issue of water scarcity and the attendant social and environmental challenges share striking similarities to approaches employed elsewhere. First, an economic philosophy has been adopted to achieve a more efficient allocation among perceived private good users, particularly in the irrigation sector (Boddington and Synnott 1989; Syme and Nancarrow 1991). Secondly, a managed or bureaucratic approach has been encouraged where there are implicit public good uses for water (Handmer et al. 1991; Syme and Nancarrow 1991). Again, this reflects the growing acceptance at the political and bureaucratic levels of the legitimacy of environmental claims on water resources (see, for example, Syme and Nancarrow 1991). Against these claims are set the views of irrigators who have expressed concern about the perceived generosity of environmental allocations and seemingly unrealistic provisions for environmental risk (Crase et al. 2000). Similar policy conundrums exist elsewhere, but what sets the context of Australian water policy apart is its geographic inheritance and unique political, economic, and social institutions. Thus, while this is a testing time for water policymakers as they attempt to balance the seemingly incongruous values of water stakeholders and grapple with the attendant social ramifications, there is also enormous potential to learn from these experiences.

Approach and Scope of the Book

The perception of the development and implementation of water policy as a social and political task contrasts starkly with prevailing approaches that emphasize the technical or engineering elements of water management. However, the preceding description of Australia's water problems highlights the necessity for examining water policy from the perspective of a number of disciplines; a challenge attempted in this book.

The motivation for employing multidisciplinary research to inform water policy stems from several sources. First, the physical dimensions of water cannot be ignored. Water is fugitive, variable in terms of quantity and quality, and difficult to control in a physical sense. Moreover, in Australia's case, the combination of government determination to develop the resource coupled with engineering zeal has created a water bureaucracy strongly influenced by engineering technocrats. The early prerogative of engineers to control water policy was encapsulated by Hugh McKinney, the first head of the NSW Public Works Department's Water Conservation Branch in 1887. Although highly opinionated, the breadth of influence of the engineering profession was encapsulated by McKinney (1887, 60) when he claimed that "[n]o one is more entitled to be heard on the general principles of water supply of a country than an engineer, who has had practical experience both of the value of good laws, and the mischief caused by bad ones." Although the influence of the engineering profession has been weakened by attention to emerging environmental ills and economic concerns relating to the management of infrastructure, the engineering perspective still plays an important part in contemporary water policy formulation. Recently, this influence has been reinvigorated by the increased enthusiasm for technical solutions such as reclaiming sewage water for re-use and desalination as a means of securing the water supplies for large urban areas including Sydney and Perth (see, for example, Frew and Noonan 2005).

Second, the role of scientists, and particularly ecologists, has been enhanced in the water policy arena. This has occurred in conjunction with the expanded community awareness of environmental degradation. Scientists have played a major role in re-shaping many of the early water reforms of the 1990s, which primarily focused on economic objectives. Now most reforms encompass a clear environmental mandate. For example, the *Living Murray* process and the ensuing debate have, in large part, been motivated by the efforts of scientists to fill important knowledge gaps about the natural riparian environment. The audience attracted by the *Wentworth Group of Concerned Scientists* during deliberations over environmental flows stands as testament to the influence now enjoyed by this profession in the water discourse.[1]

Third, in addition to the physical sciences, scholars and practitioners in the social sciences are adding significantly to contemporary water policy formulation. This approach is consistent with the growing body of work that depicts the problem of water management as having its origins in humanly devised interactions, rather than being purely technical or ecological in nature (see, for example, Ostrom 1992). On the basis of this observation, economists, sociologists, and lawyers all have important parts to play.

As noted earlier, the Australian water reforms of the 1990s were but one manifestation of the wider enthusiasm for economic reform. But economic restructuring was not unique to Australia. Massive economic transformations were occurring at the international level, witnessed by the collapse of communist regimes and the rise of market capitalism in China. At the national level, competition was increasingly embraced as the mechanism for accelerating growth and maintaining prosperity. This philosophy has also informed natural resource management and encouraged alternative ways of conceptualizing and ameliorating environmental harm. The resultant need for institutional adjustments provided work for lawyers in redesigning property rights (see, for example, McKay 2003), produced social laboratories for examining the ramifications of change (see, for instance, Syme and Nancarrow 2002), and spurred additional speculation among economists about the rationale for markets, particularly in the context of natural resources (see, for example, Crase et al. 2004).

Put simply, it is difficult to find a discipline that cannot offer a useful perspective on Australian water policy; however, the task of bringing the various perspectives into a policy consensus is not without its challenges. Mullen (1996) and others have frequently noted the incongruity between disciplines and the costs of producing meaningful multidisciplinary outcomes. Notwithstanding the extent of these problems, this book represents an attempt to provide some understanding of the breadth of water policy, the contribution that can be made by differing disciplines, and the theoretical and practical lessons that emerge from serious analysis.

Objectives of the Book

Clearly, water policy is not created in a vacuum. An understanding of the hydrological, geographical, historical, social, economic, and political perspectives is imperative to gaining an appreciation of the challenges confronting water policymakers. This is particularly the case in Australia where, as has already been noted, significant variations occur between jurisdictions and even within states; however, it is also this variety that provides opportunities for the policy analyst to conduct comparative assessments.

In order to make the task of examining the range of Australian water policies manageable, answers have been sought to five broad questions:

1. What is the magnitude and contour of water challenges in Australia?
2. What are the motivations that are driving water reform in Australia?
3. What is the ideal state of water resource management to which Australians aspire?
4. What are the key characteristics of reform, and the impediments to and consequences of the reform journey?
5. What lessons might be derived from the reform experience?

The objective of this book is to provide readers with sufficient knowledge of water policy in Australia to allow them to formulate coherent responses to these questions. In so doing, the reader is exposed to an examination of the agro-environmental challenges being dealt with in Australia. In addition, the water scarcity confronting urban communities is uncovered. In this context the nation's largest cities are struggling to achieve the balance between urban/industrial growth, meeting environmental objectives, and maintaining affinity with rural communities, many of whom have a history of generous water use in irrigation.

The wide array of political, social, and economic problems that circumscribe water allocation decisions in Australia has already provided fertile ground for scholars, analysts, and commentators. This has manifested itself in dialogue on the dilemmas and solutions to water resource scarcity. Accordingly, a review of Australia's attempts to manage water resources provides both a theoretical and a practical guide to contemporary water management.

Structure of the Book

The remainder of this book is broadly organized into three parts. In the first section a more detailed overview is provided of the context of Australian water resource policy. Comprising five chapters, this section includes treatment of the hydrological, historical, legal, and institutional dimensions of water policy.

In Chapter 2 Rebecca Letcher and Susan Powell describe the biophysical nature of water resources in Australia, including both quantity and quality aspects of groundwater and surface water resources. The spatial and temporal distribution of these resources, including their variability and reliability, and the location of resources relative to demand are also considered.

The focus in Chapter 3 is on the evolution of water policy, primarily within the context of the Murray-Darling Basin. As noted earlier, the Basin not only accounts for most of Australia's water development but also provides a valuable insight into the way water policy has been conceptualized elsewhere in Australia. Warren Musgrave reviews the policy developments in the Murray-Darling Basin and divides important episodes into three distinct eras.

In Chapter 4, Jennifer McKay scrutinizes the details of the legal framework that has evolved against the historical backdrop provided in Chapter 3. Her analysis extends to the current phases of water policy which are characterized by greater attention to the principles enshrined in Environmentally Sustainable Development. Special attention is given to the National Water Initiative and the legal challenges posed by the approach articulated in the Initiative.

The broad notions of risk and uncertainty and their distribution among water stakeholders are explored by John Quiggin in Chapter 5. Variability of supply typifies Australian water resources and is considered from a policy perspective in this chapter. John also provides some preliminary insights into the contemporary reform and coping strategies examined in the final part of the book.

Lin Crase and Brain Dollery specifically explore water policies and their relationship to property rights in Chapter 6. In addition, the modification of property rights in response to changing social norms is examined. An evaluation of the hierarchical modes of water governance and their definition in the context of Australian water management is also considered.

Having delineated the context for the formulation of Australian water policy, the second part of the book concentrates on the sectors commonly targeted by policymakers.

Jenny McLeod and George Warne describe the important contemporary issues facing the irrigation sector in Australia. These issues primarily result from 15 years of government water reforms which reflect a fundamental shift in government and community attitudes to irrigation, away from development, expansion and associated social objectives to encourage closer settlement, to a preoccupation with full cost recovery and environment protection. Jenny and George's observations are grounded in the experiences of Murray Irrigation Limited, an irrigation company in the southern Murray-Darling Basin that was established when the government-owned irrigation districts were privatized in 1995. Parallels with water policies faced by the Australian irrigation industry outside of the Basin are also drawn in this chapter.

Hydroelectricity in Australia has mainly been limited, by topography and hydrology, to Tasmania and the Snowy Mountains scheme. Nevertheless, the sector occupies an important place in the Australian water policy milieu, and Ronlyn Duncan and Aynsely Kellow examine its significance in Chapter 8. As with irrigation, hydroelectricity was originally typified by a strongly developmentalist approach which has now been replaced by a more mature rationale for water resources management. The chapter surveys this policy shift from the perspective of hydroelectricity and also examines important interactions between hydroelectricity and other sectors.

The move from a development to management approach has been underscored by recognition that conquering nature is a forlorn ambition and that establishing a rapprochement with the ecosystem that supports a natural resource is essential to its indefinite use. This is the focus adopted by Terry Hillman in Chapter 9.

A synoptic overview of urban water management in Australia is presented by Geoff Edwards in Chapter 10. He considers selected cases and highlights significant differences in approach across jurisdictions. Emanating from this comparison is an understanding of the political motivations influencing urban water policy and an appreciation of how this policy context differs from that of the agricultural sector.

The third general part of the book considers water reform more broadly and explores coping strategies being adopted in the Australian setting. Sue O'Keefe and Lin Crase use Chapter 11 to examine the information that has prompted surface and groundwater policy reforms and briefly introduce a typology of responses. These include trade scenarios and water recovery projects.

The roles of technology generally, and water re-use in particular, are considered by Stuart Khan in Chapter 12. Notwithstanding the need for behavioral and institutional adjustments highlighted in other chapters, the potential for technology to act as a vehicle for addressing some concerns is also being examined by Australian policymakers. Stuart highlights the growing interest in technological adjuncts to assist in addressing water scarcity. Observations drawn in this chapter offer a useful adjunct to the review of urban water policy presented in Chapter 10.

John Rolfe then explicitly deals with the vexing issue of water's value and the underlying hypothesis that policy reform should move water resources to yield their highest return. He explores evidence of the present impacts of water trade on agriculture in the context of both seasonal and permanent transfers. The negative environmental consequences that have arisen from trade in some instances are also detailed.

Phil Pagan uses Chapter 14 to consider the concept of adaptive management. This chapter also provides a useful complement to the risk allocation framework discussed by John Quiggin in Chapter 5. While briefly touching on the theoretical dimensions of adaptive management, Phil focuses primarily on the Australian data and experiences. The role of adaptive management in Australian water markets and the implications of the *Living Murray* process receive specific attention.

Chapter 15 explores the social and cultural dimensions to water. Geoff Syme and Blair Nancarrow expose the deficiencies in attempts to systematically account for subjective social and cultural issues in developing long term planning and performance monitoring standards.

In Chapter 16 concluding remarks are shaped around the five core questions that provide the motivation for this book. The chapter draws on the valued perspectives offered by all contributors and attempts to briefly summarize responses to these questions.

Currency Amounts

All currency amounts in the book are in Australian Dollars (AUD). As of November 2, 2007, $1 AUD = $0.92 USD.

Notes

1. The part played by the Wentworth Group in Australian water policy has not gone without attracting some criticism (see, for example, A. Watson 2003).

References

ABS (Australian Bureau of Statistics). 2002. Condition of Australia's freshwater resources. *Australian Year Book*. Canberra: Australian Bureau of Statistics. http://www.abs.gov.au/ausstats/abs@.nsf/0/3f8d7e912784b4eaca256b35007ace03?OpenDocument (accessed July 1, 2005).
ABS (Australian Bureau of Statistics). 2003. *Australian Year Book*: National Accounts. Canberra: Australian Bureau of Statistics. http://www.abs.gov.au/Ausstats/abs@.nsf/0/38a0849024af4249ca256cae00052114?OpenDocument (accessed June 9, 2005).

Australian Academy of Science. 2004. Monitoring the white death—Soil salinity. http://www.science.org.au/nova/032/032key.htm (accessed May 20, 2005).

Blakers, A. 1994. Hydroelectricity in Tasmania revisited. *Australian Journal of Environmental Management*: 110–120.

Boddington, W., and M. Synnott. 1989. *An examination of alternative methods of groundwater allocation*. Perth: Western Australian Water Resources Council.

Challen, R. 2000. *Institutions, transaction costs, and environmental policy: Institutional reform for water resources*. Cheltenham: Edward Elgar.

Clark, S. 1971. The River Murray question: Part 1—Colonial days. *Melbourne University Law Review* 8:11–40.

Coop, P., and D. Brunckhorst. 1999. Triumph of the commons: Age-old participatory practices provide lessons for institutional reform in the rural sector. *Australian Journal of Environmental Management* 6 (June): 69–77.

Crase, L., L. O'Reilly, and B. Dollery. 2000. Water markets as a vehicle for water reform: The case of New South Wales. *Australian Journal of Agricultural and Resource Economics* 44 (2): 299–322.

Crase, L., P. Pagan, and B. Dollery. 2004. Water markets as a vehicle for reforming water resource allocation in the Murray-Darling Basin. *Water Resources Research* 40: 1–10.

Culture and Recreation Portal. 2005. Australian Farms and Farming Communities. Australian Government. http://www.cultureandrecreation.gov.au/articles/farms/ (accessed September 20, 2005).

DNRM (Department of Natural Resources and Mines). 2002. *Queensland irrigation schemes—A brief history*. Brisbane: Department of Natural Resources and Mines.

DPMC (Department of Prime Minister and Cabinet). 2005. *About the national water initiative*. Canberra: Department of Prime Minister and Cabinet.

DSE (Department of Sustainability and Environment—Victoria). 2004. *Securing our water future together*. Melbourne: Department of Sustainability and Environment.

The Economist. 2003. Priceless: A survey of water. 368 (8333): 1–16.

Frew, W., and G. Noonan. 2005. Water price rise to fund desalination. *SydneyMorning Herald*, April 28. http://www.smh.com.au/news/National/water-price-rise-to-fund-desalination/2005/04/28 (accessed June 12, 2005).

Green, C. 2003. *Handbook of water economics principles and practices*. Chichester: Wiley.

Hallows, P., and D. Thompson. 1998. *The history of irrigation in Australia*. Mildura: Australian National Committee on Irrigation and Drainage.

Handmer, W., A. Dorcey, and D. Smith, eds. 1991. *Negotiating water: Conflict resolution in Australian water management*. Canberra: Centre for Resource and Environmental Studies.

Johnson, S., D. Commander, C. O'Boy, and R. Lindsay. 2005. *Groundwater investigation program in western Australia (2005 to 2020). Hydrogeological record series*. Perth: Department of Environment.

Kenyon, P., and A. Black, eds. 2001. *Small town renewal overview and case studies*. Canberra: RIRDC.

Kingsley, D. 2002. Perth water crisis looms. *News in Science ABC Science Online*, August 27. http://www.abc.net.au/science/news/stories/s659064.htm (accessed June 1, 2005).

Langford-Smith, T., and J. Rutherford. 1966. *Water and land: Two case studies in irrigation*. Canberra: Australian National University Press.

Lloyd, C. 1988. *Either drought or plenty: Water development and management in NSW*. Parramatta: Department of Water Resources NSW.

Macquarie Dictionary of Australian English. 2003. Sydney: Macquarie Dictionary.

Magarey, A. 1894–1895. Aboriginal Water Quest. *Proceedings of the Royal Geographical Society of Australasia, South Australian Branch*: 1–15.

McKay, J. 2003. New directions and national leadership in developing water policies in federations—India and Australia. Paper presented at the International Workshop on Institutional Design: India and Australia. July 2003, Beechworth, Australia.

McKinney, H. 1887. Notes on the experience of other countries in the administration of their water supplies. *Journal and Proceedings of the Royal Society of New South Wales* 21:60–73.

MDBC (Murray-Darling Basin Commission). 1998. *Managing the water resources of the Murray-Darling Basin.* Canberra: Murray-Darling Basin Commission.

MDBC (Murray-Darling Basin Commission). 2002. *The Living Murray: A discussion paper on restoring the health of the River Murray.* Canberra: Murray-Darling Basin Ministerial Council.

Mullen, J. 1996. Why economists and scientists find cooperation costly. *Review of Marketing and Agricultural Economics* 64 (2): 216–24.

Mulvaney, J., and J. Kamminga. 1999. *Prehistory of Australia.* St. Leonards, NSW: Allen and Unwin.

Ostrom, E. 1992. *Crafting institutions for self-governing irrigation systems.* San Francisco: ICS Press.

Pannell, D. J. 2005. Salinity: New knowledge with big implications. http://cyllene.uwa.edu.au/~dpannell/dp0504.htm (accessed August 21, 2005).

Randall, A. 1981. Property entitlements and pricing policies for a mature water economy. *Australian Journal of Agricultural Economics* 25 (3): 195–220.

Syme, G., and B. Nancarrow. 1991. Community analysis of water allocation. *Water Allocation for the Environment: Proceedings of an international seminar and workshop.* Ed John J. Pigram and Bruce P. Hooper, 27–29 November 1991, Armidale, NSW: Centre for Water Policy Research.

Syme, G., and B. Nancarrow. 2002. Evaluation of public involvement programs: Measuring justice and process criteria. *Water* 29(4): 18–24.

Thorne, A., R. Grun, G. Mortimer, N. Spooner, J. Simpson, M. McCulloch, L. Taylor, and D. Curnoe. 1999. Australia's oldest human remains: Age of the lake Mungo 3 skeleton. *Journal of Human Evolution* 36:591–612.

Turner, S. 2005. Statement by Simon Turner, President MCAV: Government decision on alpine grazing devastating. http://www.mcav.com.au/grazing_ban.html#2 (accessed July 1, 2005).

Watson, A. 2003. Approaches to increasing river flows. *The Australian Economic Review* 36 (2): 213–24.

Watson, B. 1990. An overview of water sector issues and initiatives in Australia. Transferability of water entitlements: An international seminar and workshop, Armidale, Centre for Water Policy Research, University of New England.

Watson, D. 2004. *Don Watson's dictionary of weasel words: Contemporary clichés, cant and, management jargon.* Sydney: Random House Australia.

CHAPTER 2

The Hydrological Setting

Rebecca Letcher and Susan Powell

IN ORDER TO CONSIDER THE WAY in which land and water management is undertaken in Australia, it is important to understand the physical
dynamics in which this occurs. As noted in Chapter 1, Australia has a unique
hydrological setting that strongly affects the success of strategies for managing land and water resources. The nuances of this setting also place special
demands on the ways that these resources can be (and have been) harnessed.
This chapter provides a brief overview of the hydrological setting of the
Australian continent, including rainfall and climate regimes as well as surface and groundwater resource quality and quantity. Much of this material
has been derived from two definitive sources on Australian water resources:
Smith (1998) and the National Land and Water Resources Audit (2005).
Hopefully this sketch of the physical aspects of Australia's water resources is
adequate to buttress an understanding of the policy analysis presented in later
chapters. The remainder of this chapter is organized into four sections. First,
we consider the Australian climate by focusing specifically on rainfall and
evaporation characteristics. Second, water availability is addressed by examining surface and groundwater resources. Third, water quality parameters are
reviewed with particular emphasis on the spatial variability of quality indicators. Finally, we offer some brief concluding remarks.

Australian Climate: Rainfall and Evaporation

Australia's rainfall varies greatly both spatially and temporally. The northern
part of Australia, covering the top of Queensland, Western Australia, and the
Northern Territory, experiences distinct wet summer and dry winter seasons.
The majority of the Australian land mass is arid, consisting of central Australia

and a large part of the central coastal area of Western Australia. The summer rainfall zone covers southeastern Queensland and the northeast of New South Wales (NSW). Further south, rainfall moves from a uniform distribution across the year (much of NSW and the southeast corner of Tasmania) to a winter rainfall zone in south Western Australia, South Australia, Victoria, and Tasmania. Relatively small areas are truly winter dominant, with wet winters and dry summers. These areas occur on the southwest coast of Western Australia and in very small patches in South Australia.

Average annual rainfall on the east coast is generally greater than that on the west coast. Average annual rainfall along the coastal fringe also generally exceeds that of inland areas, ranging from over 2,400 mm/year in parts of the tropics of Queensland and the west coast of Tasmania, to approximately 1,200–2,000 mm/year along most of the NSW, Queensland, and Northern Territory coastline. In contrast, rainfall is usually less than 200 mm/year in inland Australia. The southern tip of the coast of Western Australia also experiences relatively wet conditions (~1,000 mm/year) compared to the remainder of the state.

Temporal variability can be measured by estimating the difference between the 90th and 10th percentile of rainfall and normalizing the estimate by the median rainfall value. The resulting figure classifies variability as low where this value is less than 0.5 whereas extreme variability is assigned a value greater than 2. Areas that are classified as having high or very high variability generally occur in Central Australia, but extend across to the central coast of Western Australia. Moderate to high variability of rainfall occurs in western NSW, South Australia, western Queensland and Western Australia and the Northern Territory. Low or low-to-moderate variability occurs only at the extremes of the continent—Tasmania and the southern parts of the eastern states (extending as far as southern Queensland along the coast) or the northern tip of Western Australia, Queensland, and the Northern Territory. Variability is also correlated with low rainfall areas—that is, areas with low annual average rainfall generally have higher rainfall variability. Love (2005) uses the coefficient of variation to measure rainfall variability. He finds that the overall variability of Australian rainfall is over 17 percent, although variability in the high rainfall zone has ranged from approximately 14 to 15 percent over the last century. Other areas of Australia have a greater variability than this zone, with variability in the pastoral zone, which covers most of the interior and north of Australia, ranging from 17 percent from 1900 to 1950 to 25 percent in the period since 1950. The wheat/sheep zone, where the majority of cropping and irrigation occurs, had variability between 17 and 18 percent over these periods. In contrast, of 10 other countries examined by Love (2005) over the same period, none had rainfall variability higher than 14 percent. Moreover, most of these countries, with the exceptions being South Africa, Germany, and France, had rainfall variability of less than 10 percent.

The second key climate variable which strongly affects water availability is evapotranspiration. This is a measure of the evaporation from the earth's surface as well as transpiration of water by vegetation and is a function of solar radiation, air temperature, and humidity (NLWRA 2005). It is essentially water lost to surface and groundwater stores and so strongly affects the quantity of water available for harvesting, environmental flows, or passive uses. Potential evaporation (ET) is the amount of water that would theoretically evaporate from an open water surface.

In general, potential ET increases further north in Australia due to solar radiation and decreases in proximity to the east coast due to cloudiness (NLWRA 2005). In many areas, potential ET is greater than, or at least close to, average annual rainfall. For example, the minimum value of the potential ET scale is 800 mm/year, but annual rainfall varies to as little as 200 mm/year. Put simply, this means that in many areas of Australia surface water is likely to be evaporated or transpired within a reasonably short period after rainfall. Water stored as groundwater is likely to be the only naturally available water source in many areas for much of the time. Water lost from storages such as dams is also greater, meaning a larger volume of storage is required to achieve a given security of supply. The depth of storages and their surface area strongly affects the relative volumes of storage lost to evaporation.

In summary, Australia's climate—that is, rainfall and evaporation—are characterized by a high degree of spatial and temporal variability.

Water Availability

Rainfall variability and the relatively high potential ET mean that Australia, on average, has a very low rate of runoff by comparison with other continents. Such comparisons can be undertaken by considering runoff per unit area (mm), which stands at 1.6 mm for the Australian continent. This is well below the next lowest value of 4.8 mm in Africa. Australia accounts for over 5 percent of the world's continental land mass but generates only 1 percent of the total runoff.

The variability in rainfall and ET described earlier also means that runoff is quite variable across the continent. On average, 12 percent of Australian rainfall runs off to rivers; however, this varies considerably across the country (AWRA 2000). The largest area, comprising of much of arid Australia, accounts for only 0.4 percent of the total runoff. The Murray-Darling Basin, which accounts for approximately 40 percent of the national income derived from grazing and agriculture (MDBC 2005), generates only 6.2 percent of the total runoff. The relatively sparsely populated areas of northern Western Australia, Northern Territory, and Queensland together account for over 60 percent of Australian runoff, whereas the small mass of Tasmania produces a relatively high 11.8 percent of runoff. Moreover, runoff is commonly produced along the coastline—and therefore discharges to sea—whereas inland areas are comparatively drier.

Generally, the areas of greatest harvestable runoff do not coincide with the areas of greatest demand. Diversions from several regions (Indian Ocean, Timor Sea, Gulf of Carpentaria, Bulloo-Bancannia, and western Plateau) are insignificant in comparison with the total volume of runoff (<0.3 percent) and, on average, less than 5 percent of total Australian runoff is diverted; however, these data mask the significant hydrological stress being placed on particular basins where demand outstrips supply.

For instance, the data available on the Murray-Darling Drainage Division are illustrative of the concentrated nature of the problem. Even without accounting for unregulated diversions in NSW, over 50 percent of mean annual runoff is diverted. In contrast, the next highest level of diversion is 15 percent and occurs in the South Australian Gulf region. All other drainage divisions have less than 10 percent of their total runoff diverted, although it should be noted that data for Tasmania exclude the activities of the Hydroelectricity sector.

Surface Water

Surface water resources are generally harvested in major storages or farm dams. Additionally, the variability of rainfall and evaporation patterns means Australian storages must be designed to a larger capacity to ensure an equivalent security of supply as that experienced in countries with less variable water resources. These factors obviously affect the returns to investment in storage capacity in Australia.

The Australian Water Resource Audit (2000) found that there were 447 large dams in Australia with a combined capacity of 79,000 GL. In addition, the Audit found that there are several million farm dams which account for an estimated 9 percent of total stored water. The volume of water stored in large dams varies significantly across the states, as illustrated by Table 2-1. In general, the data show that, as of 1990, New South Wales and Tasmania had double the storage capacity of Victoria, Queensland, and Western Australia. Other states and territories, notably South Australia, had very low levels of surface water development.[1]

Significant storage capacity occurs in small pockets of Australia, but across most parts of Australia, drainage divisions have a low state of development. The populous eastern states of Australia, however, are in many cases highly or fully developed, and for much of the Murray-Darling Basin, over-developed (see NLWRA 2005). Tasmania and south Queensland also have areas of high development.

Table 2-1. *Storage Capacity (GL) in Large Dams to 1990*

State	Storage capacity (GL)
New South Wales	25,389
Victoria	12,226
Queensland	9,459
Western Australia	7,011
South Australia	267
Tasmania	24,167
Northern Territory	275
Australian Capital Territory	125

Source: IEA (1999).

Not surprisingly, there is a strong correlation between the extent of surface water development and the availability of additional surface water resources. Many areas of NSW and Victoria are either fully or over-developed such that any move to create environmental allocations in these systems, coupled with the natural variability of rainfall in these areas, is likely to impact the availability of water for irrigation.

Where storages are used to supply irrigation and town water, these supplies have the effect of reversing the seasonal pattern of flows, a theme explored in greater detail later by Hillman (Chapter 9). Many Australian rivers are either ephemeral or winter dominated in their flow pattern. Storages and associated irrigation releases create river conditions which are characterized by constant, summer-dominant flow patterns. These flows frequently comprise water which is significantly cooler than natural flows.

Groundwater

Groundwater is used for irrigation, stock, and domestic supplies, as well as for industrial and town water use. It has been estimated that Australia has 25,780 GL of groundwater of sufficient quality for potable, stock, and domestic or irrigation use that can be sustainably extracted annually (AWRA 2000). A major groundwater system underlying much of northern New South Wales, Queensland, South Australia, and part of the Northern Territory is the Great Artesian Basin (GAB). This groundwater system extends for 1.7 million km², stores a volume of 8,700,000 GL of water, and supplies 570 GL annually to grazing and mining interests (AWRA 2000). The sustainable yield of the GAB has been estimated at 600-1,200 GL/year. Many groundwater systems, particularly in Western Australia, have sustainable yields of less than 600 ML/year. Highly developed aquifers occur on the coastal fringe of south Western Australia, in the Northern Territory, on the border areas of New South Wales, Victoria, and South Australia and on the Queensland and northern New South Wales coast.

The development status of Australian aquifers can be measured by expressing use as a proportion of sustainable groundwater yield. The majority of highly developed aquifers (i.e., those over 70 percent allocated) occur in the eastern states of NSW and Queensland, with smaller areas in Victoria and Western Australia (see AWRA 2000). There is considerable overlap between groundwater resources that are over 100 percent allocated in New South Wales and surface water resources that are over-developed. This means that there is significant stress on water resources in these areas and that current production activities are close to full capacity of the water system.

Water Quality

The second major factor that affects the availability and usability of water resources is the quality of surface and groundwater resources. Water quality can be defined by its physical, chemical, biological, and aesthetic characteristics. It is strongly influenced by a range of land management practices as well as through direct pollution and management of the resource.

Water quality is generally assessed in relation to a range of specific uses. These include the capacity of the water to support biodiversity and aquatic ecosystem functioning, quality thresholds for aquaculture, human drinking water, recreation (including both primary and secondary contact activities), irrigation, stock watering, and industrial processes.

Water quality affects a diverse range of industries including tourism, fishing, and agriculture. Poor water quality increases the water treatment costs for consumptive use as well as adversely impacting aquatic ecosystems and biodiversity (AWRA 2000). For example, it has been estimated that freshwater algal blooms cost the Australian community between $180 and $240 million each year (LWRRDC 2000).

There are a range of detailed measures of water quality covering biological, physical, chemical, aesthetic, and radioactive parameters. In most instances in Australia only a small number of indicators are consistently monitored.

Many water quality characteristics are strongly correlated to river flows. Generally, high flows resulting from rainfall and runoff in a catchment result in higher turbidity, nutrients, and a range of pollutants. Salinity is often inversely related to flow. These relationships reflect the pathways by which pollutants enter stream systems, whether they are through baseflow to streams from groundwater or as overland flows. Many pollutants arrive in waterways attached to sediments and are carried in large overland flow events.

Drewry et al. (2005) analyze flow and sediment data from the Eurobodalla river catchment in southeastern NSW. Their work demonstrates a typical relationship between suspended sediment and flow for an Australian catchment. They plot suspended sediment and flow data measures over 15-minute intervals during two storm events occurring over a 20-day period. They show that as flow increases, suspended sediment concentrations also increase. As flow falls, after the storm has passed, concentrations also fall. Overall, 98 percent of the total suspended sediment load measured over a 20-day period is generated over just 2 percent of the time. This illustrates not only the correlation of flow and water quality, but also the great variability in measurable water quality over time.

Assessment of water quality on a continental scale is even more difficult due to the paucity of data. AWRA (2000) found that turbidity and nutrients were identified as the most widespread water quality issue, followed by salinity and acidity/alkalinity; however, these factors, particularly turbidity, are also the most commonly monitored.

Surface water quality guidelines are determined by water samples meeting ecological, social, and economic requirements based on protection of aquatic systems, drinking water, agricultural water, recreation, and aesthetics (NLWRA 2001). The National Land and Water Resource Audit (2001) defined major exceedances as those that occupied in more than 33 percent of the basin area. Significant exceedances were deemed to exist where they occupied greater than

5 percent but less than 33 percent of the basin area. Although data were not readily available for all jurisdictions, 75 to 80 percent of assessed river basins had significant or major exceedances of turbidity and nutrients (total nitrogen and total phosphorus). Over 50 percent also exceeded salinity levels and 30 percent exceeded acidity guidelines.

Nutrients

Nutrients are derived from many sources including sewerage treatment works, runoff from agricultural lands, natural occurrence in soils, farm and industrial effluent, and urban storm water. Phosphorus and nitrogen are the most commonly monitored nutrients. They can be attached to suspended materials such as soil or organic matter, bound in sediments, or dissolved. The main transport mechanism for phosphorus is through soil erosion. Nitrogen is also transported in dissolved form through groundwater flows into surface water systems.

The National Land and Water Resources Audit (2001) found that the dominant sources of phosphorus are hill/slope erosion in Queensland and New South Wales; gully and riverbank erosion and dissolved phosphorus arising from runoff in coastal Victoria, South Australia, Western Australia, and Tasmania; and urban point sources in some river basins (for example, 30 percent of total phosphorus load for Moreton Bay).

Nutrients are also removed from the water column through a number of processes. Major nutrient sinks are floodplain sedimentation, where sediments are deposited on the floodplain as floodwaters recede. These sediments usually contain substantial nutrient loads. In-stream and off-stream reservoirs also act as sinks for nutrients and sediments, trapping these nutrients before they are transported further downstream. Finally, for nitrogen, in-stream denitrification is also a significant sink. This is where nitrogen is exchanged between the water column, plants, and the atmosphere through biological and chemical processes.

The distribution of basins with nutrient management concerns has been well documented by NLWRA (2005). This work clearly shows the lack of available data across much of Australia. Major nutrient problems were also identified across much of NSW, Victoria, and Queensland. Only a very small number of basins with available data were assessed as having no nutrient problems. The distribution of basins with nutrient management concerns largely corresponds to the distribution of phosphorus. Nitrogen is less commonly monitored.

Turbidity

Turbidity is the presence of suspended soil particles and colloids in the water. It can be caused by mobilization of sediments from the catchment (exacerbated by land-use activities), re-suspension of sediments within river systems, or stream bank and gully erosion. Turbidity is a major water quality issue in 41 (61 percent) of the 67 basins measured in the Australian Water Resource Assessment (AWRA 2000). Turbidity is naturally high in many basins due to variable climate and streamflow as well as highly erodible soils and stream banks. The NLWRA (2001) has mapped the distribution of turbidity management issues across Australian drainage divisions. This closely resembles the distribution of nutrient management problems, reflecting the importance of sediment as a transport mechanism for nutrients. As for nutrients, turbidity is a major concern in many of the basins in eastern Australia. This work also demonstrates the relative paucity of data to understand turbidity on a national scale.

Salinity

Salinity is the presence of dissolved salts in soil and water and is a problem common to many parts of Australia. It may be caused by the presence of salt in underlying soil or bedrock, salt deposited in past marine inundation of an area, or salt particles being carried over the land surface from the ocean (DIPNR 2003). Salinity in rivers is caused by several transport mechanisms. As briefly noted by Crase in Chapter 1, changes in vegetation engendered by post-European settlement have led to deep-rooting perennial vegetation being largely replaced by annual or short rotation, shallow-rooting vegetation, and fallow areas. This has changed the water balance, which has led to increasing water tables in some areas. In addition, water has also been applied to the soil for irrigation in many areas, resulting in a surplus of water draining through to aquifer systems and increasing groundwater tables. In many areas of Australia, soils contain salts which may be carried to the surface by increasing groundwater tables, and then carried by overland flow into surface waters, or leached directly to rivers through baseflows from groundwater. Additionally, salt can be directly deposited on soils from ocean sources through wind.

The NWLRA (2005) has mapped the distribution of salinity effects across Australian drainage basins. As for nutrients, this mapping exercise demonstrates that there are significant gaps which make it difficult to assess salinity at a continental scale. On the basis of the available data, however, this work shows that salinity is a major concern in basins in Western Australia and western Victoria but is less problematic than nutrient loads in Queensland and NSW.

Overall, more than 50 percent of divertible surface water in the southwest of Western Australia is considered marginal, brackish, or saline (Schofield et al. 1988 in AATSE/IEA 1999). In the future, the main impact of salinity on utilizable water resources is likely to occur in the Murray-Darling Basin. This is particularly important in the context of the significant agricultural production in this area. There is also a marked increase in downstream salinity levels in the Basin (MDBMC 1987).

Salinity also detracts from the usability of groundwater resources. Although the sustainable yield of groundwater in some areas may be adequate to support activities such as irrigation or town water supply, the quality of this water may prohibit its use. Figure 2-1 shows the sustainable yield of groundwater resources in each jurisdiction classified by salinity level. The ACT had insufficient data to be included in this analysis. Also no data were available for the lowest salinity class in South Australia and the highest salinity class in Tasmania. Water with

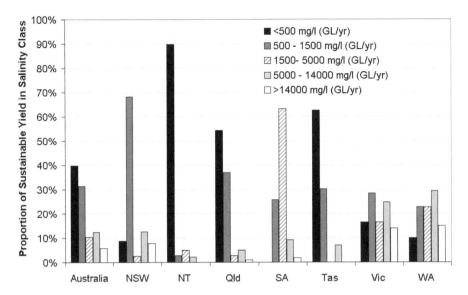

Figure 2-1. *Sustainable Yield of Groundwater Resources Classified by Salinity Level*

salinity greater than 6,000 mg/l is generally too saline for most plants and animals, although some stock and plants can use water up to 10,000 mg/l. Water between 500 mg/l and 1,500 mg/l can be used for human consumption and will support aquatic plants and animals, but it would taste very salty. Water below 500 mg/l represents good quality drinking water and is suitable for almost all animals and plants (Waterwatch South Australia 2005).

Across Australia 40 percent of groundwater sustainable yield is available from the highest quality aquifers (when measured by salinity). Less than 10 percent is from the poorest quality aquifers, although 18 percent comes from aquifers with salinity greater than 5,000 mg/l. The figure demonstrates that in the Northern Territory, Queensland, and Tasmania, the greatest proportion of sustainable yield comes from the best quality groundwater (over 50 percent in all cases). In NSW nearly 70 percent of sustainable yield is dominated by sources of less than 1,500 mg/l. There is only limited access to reasonable quality groundwater in Western Australia and Victoria, with 44 percent and 39 percent, respectively, of sustainable yield that has a salinity of greater than 5,000 mg/l. Groundwater quality in South Australia is also quite poor with over 70 percent of sustainable groundwater yield coming from aquifers with high salinity levels.

Conclusion

Australian water resource availability and quality varies greatly. This variation occurs at a spatial level, giving rise to marked differences in storage requirements across the continent. Obvious contrasts exist from north to south as well as from Central Australia to the coast. Water quality and quantity also vary temporally, with both high inter-year and intra-year variability. This creates major challenges for those seeking to manage water and to the development of institutional settings in which water users are provided with surety of supply.

Importantly, unique hydrological characteristics have given rise to idiosyncratic problems and solutions. For instance, while many Australians might aspire to productive outcomes similar to residents in Europe or America, the technological and economic demands to achieve those ends are significant in the Australian setting. Moreover, the considerable hydrological variability between and within Australian jurisdictions adds a level of institutional complexity unrivaled elsewhere. An understanding of these humanly devised reactions to the hydrological setting is the focus of remaining chapters.

Notes

1. Data for the ACT does not include Googong Dam (125 GL) which is an Australian Capital Territory-owned dam located in New South Wales.

References

AATSE/IEA. 1999. *Water and the Australian Economy.* Joint project of the Australian Academy of Technological Sciences and Engineering and the Institution of Engineers Australia.

AWRA (Australian Water Resources Assessment). 2000. *Surface water and groundwater— Availability and quality.* Australian Natural Resources Atlas V2.0. http://audit.ea.gov.au/ anra/water/docs/national/Water_Contents.html.

DIPNR (Department of Infrastructure, Planning and Natural Resources). 2003. *Water quality in the Gwydir Catchment 2001–2002.* Department of Infrastructure, Planning and Natural Resources 03_0864, Tamworth.

Drewry, J. J., L. T. H. Newham, and B. F. Croke. 2005. Estimating nutrient and sediment loads in Eurobodalla coastal catchments. *NSW Coastal Conference.* Narooma. November 2005: 8–11.

Love, G. 2005. *Impacts of climate variability on regional Australia.* Outlook 2005 Conference Proceedings, Climate Session Papers, ed. R. Nelson and G. Love. Australian Bureau of Agricultural and Resource Economics (ABARE).

LWRRDC (Land and Water Resources Research and Development Corporation). 2000. Cost of Algal Blooms, Occasional Paper 26/99, Canberra: Land and Water Resources Research and Development Corporation.

MDBC (Murray Darling Basin Commission). 2005. About the Basin. Murray-Darling Basin Commission website. http://www.mdbc.gov.au/naturalresources/about/about_basin. htm (accessed November 28, 2005).

MDBMC (Murray-Darling Basin Ministerial Council). 1987. *Murray-Darling Basin Environmental Resources Study.* Canberra: Murray-Darling Basin Ministerial Council.

NLWRA (National Land and Water Resources Audit). 2001. Australian Agriculture Assessment 2001. National Land and Water Resources Audit c/o Land & Water Australia on behalf of the Commonwealth of Australia.

NLWRA (National Land and Water Resources Audit). 2005. National Heritage Trust, Australian Natural Resource Atlas. http://audit.ea.gov.au/ANRA/atlas_home.cfm (accessed October 31, 2005).

Schofield, N. J., J. K. Ruprecht, and I. C. Loh. 1988. The Impact of Agricultural Development on the Salinity of Surface Water Resources of South-West Western Australia, Report No. WS 27. Leederville: Water Authority of Western Australia.

Smith, D. I. 1998. *Water in Australia.* Melbourne: Oxford University Press.

Waterwatch South Australia. 2005. *Salt watch.* http://www.sa.waterwatch.org.au/sw_salinity .htm (accessed December 1, 2005).

CHAPTER 3

Historical Development of Water Resources in Australia

Irrigation Policy in the Murray-Darling Basin

Warren Musgrave

T HE PURPOSE OF THIS CHAPTER IS to provide an historical setting for the analysis of contemporary Australian water policy presented
in other parts of this book. The chapter traces the historical events between
European settlement and the 1960s and offers a brief overview of later policy
developments between 1960 and 1990. This latter period is then explored in
greater detail by Jennifer McKay in the following chapter, and collectively these
chapters provide a basis for wider discussion and contrast with the current policy
agenda. The focus of this chapter is primarily on irrigated agriculture, given its
paramount status in early water resource development.

Space prevents treatment of the topic state by state.[1] Instead, focus is on the
Murray-Darling Basin, where much of Australian irrigation development is
concentrated. As has been noted in the two preceding chapters, the Murray-
Darling Basin is also critical from an economic and policy perspective, although
its hydrology is not typical of all basins in Australia. This chapter specifically
examines events in New South Wales and Victoria, the states in which the
principle extractive users of the waters of the Basin are located, and in which
the bulk of irrigation infrastructure is found.[2] Policy development and change
in the other states tended to mirror that in New South Wales and Victoria, but
with differences, particularly in allocation policies for extractive use.

Water policy in the Murray-Darling Basin (the Basin) is an amalgam of policies of New South Wales, Queensland, South Australia, and Victoria and, to an
increasing extent, of the Commonwealth or federal government. This reflects
the constitutional responsibility of the states for the natural resources within
their boundaries, along with the financial leverage of the Commonwealth as a
result of its taxing powers. The Commonwealth's role also reflects its willingness to use its constitutional powers over defense, foreign affairs, and trade to
intervene in natural resource and environmental management.

The history is divided into three periods. The first was an early establishment phase, following European settlement in 1788, when important attitudes were developed and were enshrined in seminal legislation. This was a phase of almost solely state activity, much of it before federation in 1901. The second was a development phase, marked by principally government-sponsored extensive dam building and irrigation development. Here, the Commonwealth was, at least initially, a relatively passive (financial) partner. The third is a reform phase, characterized by significant changes in attitudes toward irrigation and a growing appreciation of the limitations of past river management practices. This third stage is incomplete and has seen significant changes in legislation. It has also seen the Commonwealth playing a much more active leadership role.

The first phase started in the late nineteenth century and extended into the early twentieth. The second phase covers most of the sixty years until 1980, whereupon the third phase commenced. Discussion of the phases is preceded by consideration of two inter-related matters, which have exerted a profound influence on the development of water policy in Australia. One is the relatively great instability of Australian rainfall and streamflow. The other is the inadequacy of the common law riparian doctrine as a basis for the rights of access and use of the water in Australian watercourses. Understanding of these two points helps to develop an appreciation of the drivers of water policy development in Australia. A thesis of the chapter is that, after more than 200 years, policy has not yet fully come to grips with the implications of these two matters, particularly the first.

The Uniqueness of Australian Hydrology[3]

The British settlers brought with them at least two seriously flawed expectations as far as water is concerned. First, they expected, quite reasonably, that Australian hydrology would be similar to that of northern hemisphere countries. Second, they expected riparianism to be an adequate basis for the exploitation of water in the colony. For much of mainland Australia, these expectations were not warranted. Consequently, the assumptions the colonists made as to the appropriate institutions and works for the extraction and use of water were an inadequate and a troubled basis for the development of sound water policy over most of the nineteenth century. This was particularly true of the Murray-Darling Basin.

Australia is generally regarded as a dry continent. Although not an unreasonable observation, a more correct description is that much of the continent's water is in the wrong place, or arrives at the wrong time (Smith 1998). Even a cursory review of Letcher and Powell's earlier chapter supports this view. The highly episodic and stochastic character of Australia's rainfall has also been described by McMahon et al. (1992).

This extreme variability has profound implications for the economics of water resource development. In brief, it means that, to be viable, any level of development requires considerably greater provision for storage than is the general experience in the northern hemisphere. Smith (1998) claims that, for a given level of supply security, Australian dam storage capacities need to be twice that of the world mean and six times that of Europe. This has serious consequences for the economics of water resource development. A further implication is that, for a given level of supply security, the disruption of natural river functions, following the construction of storages and other infrastructure, is also likely to be much greater in Australia.

The Inadequacy of Riparianism

The riparian doctrine gives landholders conditional rights to access and use water contiguous with and adjoining their land (Tisdell et al. 2000). The doctrine, in effect, allows landholders to do what they wish with their (riparian) water, as long as they do not unreasonably interfere with other landholders by such use (Bates 2001). Experience has shown it to be more applicable to a situation of relatively more stable and plentiful supply than to situations such as those of Australia and the western United States (Davis 1968).

Rights under the doctrine provide a rather fragile basis for the more intense competition experienced in such relatively arid (and unstable) circumstances. This became apparent as the American West was settled, leading to the adoption of the prior appropriation, or "first in use first in right" doctrine in most western jurisdictions (Davis 1968). This doctrine provided an institutional basis for the resolution of conflicts over water, but was the source of much litigation. This contributed to rejection of the doctrine by the Australian reformers in the late nineteenth century (Powell 1989).

According to Paterson (1987), the riparian doctrine impeded the development of mining, agriculture, and towns during the nineteenth century. Such impediments included problems with appropriating land titles to accommodate catchments, storages, and infrastructure for urban development; insecure rights as a basis for raising capital for investment in such structures; obstacles placed in the way of closer settlement by the practices of the squatters, such as *peacocking* and *dummying*; and the difficulties in the control of water created by the gold rushes (Tisdell et al. 2000).

In a more general sense, and with the wisdom of hindsight, riparianism is inadequate, not just as a basis for the management of competition between productive users of water, but because it also falls far short of the provision of a sound basis for the management of the multiple and joint uses (including environmental services) of the resource. Although the initiatives of the first phase of policy development addressed the former problem, it was a century later before these wider issues were addressed.

We turn now to an outline of the policy phases identified earlier.

The First Phase[4]

Water was a major cause of unhappiness, disputation, and administrative action from the beginning of settlement in southeastern Australia (Lloyd

1988, Powell 1989, Powell 1991). Almost a century passed, however, before decisive and fundamental action was taken to address the causes of these problems, first in Victoria and subsequently in other Basin states. Smith (1998) ascribed the Victorian initiatives to an accumulation of the pressures discussed earlier, together with the climatic determinism of the 1877–1881 drought. The initiatives were taken, despite an inquiry in 1882, which concluded that irrigation would not be profitable on the Murray (Gordon and Black 1882–1883).

Water became a significant political issue and the result was the passage of a number of pieces of legislation, starting with the Water and Conservation Act of 1880 and including the important Irrigation Act of 1886. The former Act reduced the riparian rights of individuals in the interest of the state and allowed for the establishment, on local initiative, of local trusts to construct, control, and finance works. The latter Act is generally accepted as the seminal piece of irrigation legislation in Australia. It:

- exclusively vested in the state the right to the use of, flow, and the control of water in any watercourse;
- subordinated the rights of the individual in that private riparian rights could not compromise the cardinal rights of the state; and
- highlighted the need for the rights of the state and the individual to be fully defined (Mulligan and Pigram 1989).

The architect of the 1886 Irrigation Act, Alfred Deakin, argued that, without irrigation, the population of the northern plains would "be swept away, and the land must go back simply to sheep-farming" (quoted by Tyrrell 1999, 127). Deakin held that the results of Californian experience with irrigation could be repeated in Victoria. Irrigation, he claimed, would increase the value of farmland and make it desirable for families to subdivide their farms and plant orchards and vineyards that would provide a good living on intensively worked properties of as little as 40 or even 10 acres. Deakin's proposal that the colonial government play an active role in the building of dams and distribution channels was justified by the desirability of making water available to farmers over the entire northern plains, not just to those close to rivers and creeks.

A further arm of irrigation policy was the encouragement given by the Victorian government to the Chaffey Brothers to establish an irrigation colony on the River Murray at Mildura in 1887. The Canadian-born Chaffeys had established profitable irrigation colonies in California and were keen to repeat that success on a larger scale. The Victorian government's offer of a land grant was similarly matched in South Australia, and in 1887 the Chaffeys laid out another colony at Renmark, also along the Murray. The Chaffey's track record in California had been impressive, but in 1894 they filed for their bankruptcy for Australian enterprise.

The reasons for the problems of these early settlements also lay behind the difficulties experienced in later times by irrigators across the Basin. These reasons included:

- farmers' attitudes to irrigation;
- purchase of irrigation blocks by speculators;
- the pricing structures;
- engineering inadequacies; and
- inattention to marketing.

Most farmers wanted irrigation water to be available as an insurance against drought and wanted to continue the land-extensive methods of wheat/sheep farming that they knew best. In addition, many chose not to build their farming routines around the regular use of irrigation water because they feared that heavy charges would be levied for its use. The 1896 Victorian Royal Commission on Water Supply found that many farmers were only enthusiastic about irrigation because they believed it would increase land values and enable them to sell out at higher prices and move on to new land. The Chaffeys, and the other Victorian irrigation trusts, were victims of this lukewarm enthusiasm for intensive farming: most of the land subdivided at Mildura was purchased by speculators or remained vacant. Farmers were not obliged to use and pay for the water that passed through their properties. In any case, engineering problems discouraged the regular use of irrigation water: The Chaffey's pumps were expensive to run and burrowing yabbies resulted in seepage from the ditches. In other irrigation areas, the poor quality of dams made supplies unreliable. Finally, the Chaffeys did not pay much attention to the issue of the potential market for irrigated produce in Victoria, which was much smaller than in California.

Subsequently, the Water Act of 1905 strengthened the provisions of the 1886 Act. It also provided for the replacement of the trusts by a central agency responsible for irrigation and water supply and nationalized the beds and banks of all watercourses (Powell 1989).

Thus, the rights to water were nationalized and riparianism swept aside. Unfortunately, as will be seen later, although this represented an important step in the attack on the difficulties created by riparianism and a repudiation of alternatives, such as the prior appropriation doctrine of the western United States,[5] it was only the start of what was needed if water was to be managed in a sustainable way.

By the start of the twentieth century there was a general consensus among observers of Victoria's irrigation systems that the policy of making local trusts responsible for decisions about the availability and price of water had been a failure. The trusts ran into difficulty—their infrastructure was not maintained and they failed to repay loans to the state or to pay for bulk water deliveries from state-owned reservoirs. William H. Hall, an expert employed by the government of the Cape Colony (part of modern South Africa) to report on irrigation around the world, found that, in Australia, the policies of the Deakin era were flawed by "inadequate government supervision," which meant that the public sector committed funds to irrigation development "without assuming financial and technical supervisions to ensure that the expenditures were justified" (Tyrrell 1999, 150–151).

These events failed, however, to lessen the prevailing enthusiasm, which saw irrigation as providing a way of reducing climatic risk, and of providing a basis for colonial wealth and settlement of the hinterland through agricultural development and closer settlement. The desirability of irrigation and its potential for success was almost considered to be self-evident.

The Water Act of 1905 dealt with these criticisms. It abolished the irrigation trusts and paid their debts from consolidated revenue. A new statutory corporation, the State Rivers and Water Supply Commission (SRWSC), was set up to manage the state's water resources. In 1907, an American, Elwood Mead, who was "probably the country's leading authority on irrigation" (Reisner 1993, *109*), was appointed to head the Commission. Mead argued that although Deakin had seen the main task for the public sector as that of making water available, insufficient attention had been given to educating farmers and converting them to small-scale intensive cultivation.

Under Mead's leadership, the SRWSC came to be perceived as having a remit for the construction of storages, infrastructure, and ultimately, irrigation schemes themselves. Despite the failure of the trusts, the desirability of such activity was not questioned and the task was seen as being essentially technical and calling for the skills of engineers. Not surprisingly, the relevant public authorities were therefore dominated by an engineering and developmental culture, as noted by Crase in Chapter 1.

Emulation by Other States

The other states of the Basin followed Victoria's lead. That is, they all eventually vested control of water in the state and created bureaucracies to manage rural water development.

In New South Wales, experiences with water during the nineteenth century were similar to those of Victoria (Lloyd 1988),[6] although social attitudes toward irrigation development in the last two decades of the nineteenth century lacked the urgency seen in Victoria. Droughts toward the end of the century, however, stimulated a similar urge for legislative action.

In 1885 the Lyne Royal Commission advocated the expansion of irrigation along the Murray and Murrumbidgee Rivers and included the portentous suggestion of the diversion of water from the Snowy River to the Murrumbidgee (NSW Parliament 1885). In 1896, the Water Rights Act was passed with the intention of legitimizing private irrigation initiatives along the Murray and Murrumbidgee. It gave control of water to the state and provided for the licensing of the extraction of water for irrigation. In this fashion, and ten years later, New South Wales chose to copy the actions of Victoria.

The construction of the Burrinjuck Dam on the Murrumbidgee, and of distribution canals for irrigation, started in 1906. The Murrumbidgee Resumption Act and the Murrumbidgee Irrigation Act of 1910 authorized the purchase of 1.6 million acres of land on the north side of the river, below Narrandera. Elwood Mead was employed by the New South Wales government to help plan the scheme. The land was subdivided into irrigable holdings of between two and fifty acres, and roads, distribution channels, and drains were surveyed and constructed.

The Water Act of 1912 provided a comprehensive base for irrigation development for the rest of the twentieth century. This Act cemented the demise in New South Wales of the common law principles of water use and " . . . inaugurated a system of private water exploitation under public license whose essentials still apply today" (Lloyd 1988, *124*).

The River Murray Commission

Institutional arrangements for the management of the Murray were among the contentious issues during the negotiations over federation, because the river was an important watercourse impinging on three of the Basin jurisdictions, the upstream states, New South Wales and Victoria, and the downstream state, South Australia. Initially, debate was about navigation but, before it was resolved, management for extractive purposes became more important. The resulting compromise was a provision in the constitution for the Commonwealth to control navigation and shipping, and the inclusion of a guarantee of the reasonable uses of water by the states for conservation and water supply.

Agreement to balance these constitutional rights in the Murray could not be achieved by the time of federation. In fact, it took another fourteen years before the River Murray Agreement was concluded between the three states and the Commonwealth (Doyle and Kellow 1995). By this time, South Australia was more concerned with guarantees of supplies for extractive uses than in-flows for navigation.

The Agreement was enacted in 1915. It provided for equal sharing between New South Wales and Victoria of the flow at Albury, with each state retaining control of its tributaries below that point. It also guaranteed a minimum entitlement for South Australia. The River Murray Commission was established in 1917 to supervise the construction and operation of the regulatory facilities specified in the Agreement.

Overview of Phase I

The important features of the reforms of the late nineteenth and early twentieth centuries were the substitution of control by the state for riparianism, the institution of the right to use water under license, the authorization of loans by the state to failed irrigation entities, and the establishment of bureaucracies through which the states could exert their control over the resource. Unfortunately, as will be seen later, although this represented an important step in the attack on the difficulties created by riparianism and a repudiation of alternatives, such as the prior appropriation doctrine, it was only a start on what was needed if water was to be managed in a sustainable way. It did, however, lay a foundation for the enhancement of agricultural output, and the accelerated settlement of the hinterland through the expansion of state-controlled irrigation.

The clarification of rights involved in the reforms is notable in that it was solely concerned with solving the problems that related to the extractive use of water for economically productive purposes. Other uses of the resource, including amenity and environmental services, remained as neglected as they would have been under riparianism. Indeed, if anything, the provision of these services was worse because the reforms made the facilitation of extractive use possible.

Phase 2—The March of Irrigation

Irrigation Expansion

Having vested control of water, created the necessary bureaucratic agencies, and having the necessary political will to proceed, the Basin states were ready to construct the storages and infrastructure to enable the establishment of substantial areas of government-sponsored irrigation farming. There followed, for most of the twentieth century, a period of unquestioned irrigation development, dominated by engineering objectives that were, as Ward (2000) notes, of large scale but narrow scope. Ward (2000, 27) describes this extensive involvement of the state governments as follows:

> The deployment of this grand scheme received broad political and commensurate financial support and was facilitated by a well-established engineering hierarchy, responsible for the conceptualisation, planning and construction of dams and reticulated supply, drainage and sewerage systems. Additionally, the statutory authorities responsible for supplying rural water progressively controlled the pattern of rural settlement, inclusive of farm size and crop types. The agency objectives and tasks, whilst large in magnitude and scale, were narrow in scope and comprehensively specified. With minimum political distraction, the achievement of specific hydraulic and engineering objectives was vigorously executed with high levels of technical expertise and utility—there was no legislated obligation to consider external consequences, and the subsequent metric of rural water development was couched in engineering terms and measured accordingly. Although punctuated by the Depression and two World Wars, the pace of water development, particularly rural irrigation schemes, has continued unabated over the 100-year period initiated by Deakin's Irrigation Act of 1886.

As indicated, having rejected the possibility of allowing the establishment of private monopolies in the supply of bulk water, the states established public monopolies instead. The Water Conservation and Irrigation Board (WCIB) was established in New South Wales in 1896 and the State Rivers and Water Supply Commission (SRWSC) in Victoria in 1906. The remit of both bodies was the promotion and development of irrigation.

Smith (1998) refers to the steadfast and resolute pursuit of this mission by both bodies, and to the considerable power and influence they wielded. They

were aided in these respects by the considerable political and community support for irrigation and closer settlement. Equivalent bodies were eventually
established in Queensland and South Australia. All of them became involved in
not only the construction of storages and distributional infrastructure, but also
in the development and control of irrigation areas themselves. Not surprisingly,
close relationships between the irrigators and the relevant irrigation authority
were established. The resulting partnerships came to function as potent political
forces that exerted considerable influence over the direction of irrigation policy.
Hindsight shows that this influence produced results not always in the interest
of society at large.

In Victoria, Goulburn Weir (25 GL) was completed in 1889, Waranga
(411GL) in 1905 and finally, the first stage of Eildon (final capacity 3390 GL) in
1927. In New South Wales, irrigation water became available from Burrinjuck
Dam in 1912, thereby enabling establishment of the Murrumbidgee Irrigation
Areas (MIA). Establishment of the River Murray Commission provided the
basis for the development of irrigation, with Commonwealth financial involvement, along the River Murray. The first stage of the Hume Dam (final capacity
3038 GL) was completed in 1936 (Smith 1998).

Although the stage may have been set at the beginning of the twentieth century for a start on the desired expansion of irrigation, the ongoing debate over
the division of the waters of the Murray imposed a limit to ambition. Consequently, the initial steps were taken on tributaries to that stream: the Goulburn
in Victoria and the Murrumbidgee in New South Wales.

Irrigation development consisted predominantly of closer settlement schemes,
often involving returned servicemen. It was based on an ill-founded belief in the
virtues of small-scale farming.

Financial problems soon manifested themselves and concessions were consequently made to irrigators. In the case of the MIA, the New South Wales
government hoped to avoid the early problems experienced in Victoria and
South Australia by selling the land resumed for irrigation to the settlers, at
prices enhanced by the prospect of irrigation. The proceeds of these sales were
intended to pay the costs of the government investment in the scheme. In the
event, collection of payment for either land or water rates was not possible and,
in 1914 and 1916, legislation was passed suspending payments. This experience
was repeated in 1919 when soldier settlement occurred in the area. From this
point on, all pretence was effectively abandoned that irrigation would be profitable, after the costs of the provision of water had been met (Lloyd 1988).

Nonetheless, this did not stem the irrigation tide. Until the 1980s, storage
capacity was increased and new storages built, while the area of irrigable land
grew. This persistence, in the face of accumulating evidence of the fundamental economic non-viability of irrigation, is remarkable in itself. At some time
the tide could be expected to turn. And turn it did: but not for several decades
after World War II. In the meantime, two changes played an important role in
maintaining the political commitment to irrigation development in the face of
accumulating doubts about its economic viability: Rice became an important

and profitable crop in the southwestern irrigation districts of New South Wales, and the Snowy Mountains Scheme was constructed.

Following successful trials in the MIA in 1922–1923, rice cropping was rapidly accepted in the economically stressed region. In 1940, it was reported to be the most profitable crop in the area (Wilkinson 1997). Cultivation of the crop expanded to other areas in the south of the state, particularly during World War II. By 1984, the Minister for Agriculture declared that, in the southwest of the state, rice had become "the backbone of the system of irrigation areas and districts" (Wilkinson 1997). In 1981/1982, 64 percent of irrigation water in southern New South Wales Irrigation Districts was used on rice (Wilkinson 1997).

Lloyd (1988, *214*) refers to the importance of this new, profitable crop and to the creation of larger farms that resulted from the consolidations and rationalizations following the economic difficulties of the early days in the MIA. He also refers to the problems arising from the spread of the crop to the irrigation districts that lacked drainage (Lloyd 1988, *248*). Indeed, this was also a problem in the drained irrigation areas. Rice was cause for concern, as waterlogging and salinization problems started to manifest themselves throughout the irrigation areas and districts, though Lloyd points out that the crop was not solely to blame for these emerging environmental problems. He states that some waterlogging and salinization were to be expected, even if the crop was not grown (Lloyd 1988, *288*).

Diversion of the waters of the Snowy River, for a range of possible purposes, including supplying Sydney, had been suggested since the mid-1800s (Lloyd 1988). The idea gained fresh momentum in the 1940s, when New South Wales and Victoria proposed conflicting schemes; one for diversion to the Murrumbidgee and the other to the Murray. Victoria also showed greater interest in the potential of such a scheme to yield electricity.

The Commonwealth became actively involved when, in 1948, it established a technical committee, the Commonwealth and States Snowy River Committee, to look further into the matter. With Prime Minister Chifley showing enthusiasm for the Scheme, the Committee recommended a compromise plan for a diversion scheme which could generate 2,820,000 KW of electricity and provide an average of 2,300,000 acre feet of water each year for irrigation. The water was to be divided between the two states. The recommendation was accepted and, after twenty-five years, the Scheme was completed in 1974, at an estimated cost of $819 million in 1974, or around $6 billion today.

Apart from its engineering virtues, which are considerable, the process whereby the Scheme came into being was, as with the River Murray Agreement, an example of the potential for the Commonwealth to provide leadership in the resolution of conflict between the states over-boundary and trans-boundary rivers. The Commonwealth not only used its powers of persuasion, but also employed its coercive ability through its defense powers under the constitution to declare that the electricity to be produced by the Scheme was essential to the nation in wartime (Lloyd 1988). It also used its financial muscle by financing the construction of the Scheme. The states were to repay the loan from the proceeds of the sale of electricity produced by the Scheme: An arrangement amounting to the provision of a subsidy for the irrigation water from the Scheme.

The Scheme was not without controversy during its development and immediately after its completion. Davidson (1974) commented on the unsatisfactory results of the Scheme. He also reported an estimate by McColl that the same quantity of electricity could have been produced more cheaply from alternative sources. In addition, he refers to his own calculations that showed a meager return from Snowy irrigation water after allowing for all costs.[7]

The Ebb of the Irrigation Tide

Investment in storages and irrigation infrastructure continued at even higher levels after World War II. Important changes in community attitudes were, however, appearing. These first changed the nature of government involvement and then led to questioning of the desirability of continued irrigation development. The latter resulted in the virtual cessation of the construction of publicly funded storages in the Basin.

The 1960s saw a decline in support for closer settlement, as realization spread that Australia's comparative advantage lay in broad acre farming, not in the establishment of a small farm yeomanry as envisaged in the late nineteenth and early twentieth centuries. This realization applied to irrigation as well as to dry land production. Further, there was a growing appreciation that closer settlement was also an inefficient tool for the redistribution of wealth and the pursuit of social justice.

This change in attitude reinforced concerns about the fiscal burden of continued public sector development of irrigation schemes. As a result, government involvement in such schemes ceased, though its commitment to the construction of storages did not. In addition, there was a burst of smaller-scale, private dam construction in northern New South Wales in the mid-1960s to late 1970s to support expansion of the cotton industry. This expansion of storage ran in the face of the emerging sentiment against the provision of irrigation works, within the Basin and in Australia generally.

Within the space of a few years, cotton had become a major irrigated crop in the Basin. Whereas many cotton farms were family businesses, a number were owned by corporate entities. So, the tradition of smallholder, irrigated agriculture gave way to large-scale farming, in which incorporated entities played a major part, at least in New South Wales and Queensland.

From the outset, development of irrigation in Australia had its Cassandras. But it was not until the late twentieth century that their arguments were able to blunt the enthusiasm of the wider community for the romance of making the desert bloom, and the belief that the development it represented was in the overall national interest. Prior to the 1960s, the developers reigned supreme and their high priests were the leaders of the state water agencies—engineers all. That irrigation was in the overall national interest was axiomatic; and the need for critical, including economic, analysis was not considered necessary. In the battle between the economists and the pro-irrigation forces in the 1960s, the arguments of the former were, initially, dismissed by the latter with magisterial contempt.

Powell (1989) places the beginning of the turn of the tide at a symposium on water resources sponsored by the prime minister and UNESCO in September 1963, with a paper by economist Keith Campbell that sounded a call for the questioning of the previously unquestionable and pleaded for a greater input from economics in the assessment of irrigation projects. Further scrutiny of irrigation policy by academic economists followed, chiefly from Bruce Davidson. He mounted a trenchant attack on the development of irrigation in his book *Australia Wet or Dry? The Physical and Economic Limits to the Expansion of Irrigation* (1969). Ward (2000) summarizes the argument of Davidson and others as follows:

> Davidson . . . criticised the level of government expenditures on irrigation schemes, based on a thesis that drought proofing and the irrigation solution were fundamentally ill-founded and misconceived. The extant competitive advantage for Australian agriculture is founded on a high ratio of naturally well-watered land per capita. Successful agricultural enterprise was predicated on the utilisation of large tracts of cheap land, the use of low levels of labour and the production of a relatively durable export commodity. Irrigation as posited by Davidson was the antithesis of a successful Australian farming system predicated on that natural advantage. Irrigation required smaller parcels of land and was labour intensive. Davidson's examination of the accounting detail of irrigated farming budgets indicated a bleak picture for individual operators and that extensive irrigation development was economically irresponsible. (Ward 2000, *28*)

With the passage of time, the logic of the economists' case became accepted and the wider community came to doubt the value of further dam building and subsidization of irrigation water supply. This view was strengthened by the suggestion that Australia had become a mature water economy (Watson and Rose 1980, Randall 1981). To these arguments were added a number of emerging concerns about the environmental consequences of past, let alone future, irrigation development. These concerns related to the degradation and sustainability of existing levels of water use, along with questions of water quality. Clearly the institutions created in the nineteenth century for development would no longer be appropriate. Fundamental reform would be necessary. Some steps in this direction had already been taken, but the 1990s saw the commencement of explicit and formal action to achieve the integrated approach advocated by W. Watson (1990) and others.

Summary of Phase 2

The first six decades of the twentieth century saw the spread of irrigation, mainly in the Basin, in a burst of nation building, which had the virtually unquestioning support of the whole community. The public irrigation schemes, which dominated this development, were major engineering and administrative achievements. They also represented considerable struggle and sacrifice on the part of the settlers of the predominantly small farms established as a result of the prevalent closer settlement philosophy. In the 1960s, enthusiasm for this type of

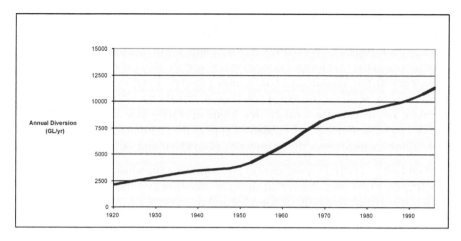

Figure 3-1. *Diversions for Consumptive Uses in the Murray-Darling Basin*

irrigation development waned, due to shifting political priorities and the abandonment of closer settlement, fiscal stringency, and mounting questioning of the economic desirability of such endeavors.

Government involvement in the construction of storages continued and the additional water was used to intensify irrigation in existing irrigation areas and districts, or to enable the growth of private irrigation, mainly for cotton in northern New South Wales and southern Queensland. As Figure 3.1 shows, the level of diversion of water for consumptive uses in the Basin continued to grow, despite the passing of government involvement in irrigation schemes and storage construction. At the end of the century, the Basin had been bequeathed a substantial irrigation industry, a large inventory of storages, and a widespread, but in places somewhat decayed, infrastructure. It also inherited an alarming level of land and water degradation, the nature of which was only dimly perceived prior to 1990.

Phase 3—The Late Twentieth Century Reforms

Irrigation development in the Basin continued at an increased rate after World War II. Forces to retard it were more than balanced by the continued advocacy of expansion. The former included the abandonment of closer settlement, fiscal pressures on the state governments, and increasing questioning of the economic efficiency of irrigation. The continued advocacy of irrigation came from the irrigation lobby, including existing and potential irrigators (particularly cotton growers), the water management bureaucracy, and a community which was reluctant to abandon its yearning to see the desert bloom.

Eventually, however, the retarding forces started to prevail. Initially, policies such as volumetric allocation and transferable water entitlements, which encouraged greater flexibility and more productive water use, were put in place. At the same time, the insistence of the Commonwealth on requests for financial assistance and accompanying benefit-cost studies, helped the spread of informed questioning of the economic merits of irrigation development. These developments were accompanied by changes in the charter of the water agencies from development to the commercial delivery of water services.

In the 1980s, fiscal pressures saw governments trying to recover a greater proportion of the costs of water supply from users. Finally, in the 1980s, and particularly in the 1990s, environmental issues, and concerns over the sustainable use of the waters of the Basin, joined criticism of the efficiency of irrigation, and of the equity of new, subsidized development.

Conclusion

The eventual acceptance of the inadequacies of the riparian doctrine in the late nineteenth century resulted in reforms which, although robust, were strongly weighted in favor of development and expanded extractive use. The resulting expansion of storages and of regulated flows eventually proved to be inconsistent with the sustainable use of the resource. Indeed, in addition to water logging and salinization, the stabilization of flows for irrigation in regulated streams was not in the interest of riverine ecosystems, which flourished in a much more irregular flow regime. Attacking these problems requires recognition of the need to allow for the environmental services of stream flows. Incorporation of this recognition in the CoAG framework, along with the use of economic instruments calculated to encourage more efficient and parsimonious water use, are therefore significant developments and are described in greater detail by McKay and Quiggin in the following chapters. The challenge for government is to successfully implement this contemporary framework against a history which encouraged, and which was itself encouraged by, a different conceptualization of the resource. Reviews of progress in this respect, and of challenges for the future, are the subject of other chapters in this book.

Notes

1. A review of irrigation development in all the Australian states is contained in Hallows and Thompson (1998).

2. Histories of water development in Queensland and South Australia can be found in Powell (1991) for the former and Hammerton (1986) for the latter.

3. This section has benefited from the review material in Tisdell et al. (2000).

4. Parts of the description of the Victorian experience in this section have benefited significantly from the material written by Lionel Frost in the preparation of Reeve et al. (2002).

5. Alfred Deakin was particularly concerned, on the basis of his study of the Californian experience, by the potential the riparian doctrine provided for the monopolization, by private interests, of bulk water services.

6. There were some successful private irrigation ventures, notably those of McCaughey who used water from the Murrumbidgee for irrigation purposes. By the end of the nineteenth century, McCaughey had about sixty miles of irrigation channels on his properties (Wilkinson 1997). N. A. Gatenby also conducted successful irrigation experiments on his property *Jemalong* on the Lachlan River from the 1890s (*NSW Agriculture Gazette*, 1903: *385–399*).

7. In the event, the Coleambally Irrigation Area, which was established to use the Snowy water in New South Wales, was never completed. Further, in order to enable them to become established, the new farmers were permitted to grow rice, but on an interim basis. Today, over thirty years later, the Coleambally district continues to be a major area of rice production.

References

Bates, G. 2001. Processes and institutional arrangements for resource and environmental management: Legal perspectives. In *Processes and institutions for resource and environmental management: Australian experiences, Final Report to Land and Water Australia,* edited by S. Dovers and S. Wildriver. Canberra: Centre for Resources and Environmental Studies, Australian National University, CD-ROM.

Davidson, B.R. 1969. *Australia wet or dry: The physical and economic limits to the expansion of irrigation.* Melbourne: Melbourne University Press.

Davidson, B. R. 1974. Irrigation economics. In *The Murray waters: Man nature and the river system,* edited by H. J. Frith and G. Sawer. Sydney: Angus and Robertson.

Davis, P. 1968. Australian and American water allocation systems compared. *Boston College Industrial and Commercial Law Review* 9: 647.

Doyle, T., and A. Kellow. 1995. *Environmental politics and policy making in Australia.* Melbourne: MacMillan.

Gordon, G. A., and A. Black. 1882–1883. *Supply of Water to the Northern Plains. Irrigation.* Parliamentary Papers (Victoria), Vol. 3, Paper No. 74, 1229–1240.

Hallows, P., and D. Thompson. 1988. *The History of Irrigation in Australia.* Mildura: Australia.

Hammerton, M. 1986. *Water South Australia: A history of the engineering and water supply department.* Adelaide: Wakefield Press.

Lloyd, C. J. 1988. *Either drought or plenty: Water development and management in New South Wales.* Sydney: Department of Water Resources.

McMahon, T., B. Finlayson, A. Haines, and R. Srikanthan. 1992. *Global runoff continental comparisons of annual flows and peak discharge.* Cremlingen-Destedt, West Germany: Catena Verlag.

Mulligan, H. K., and Pigram, J. J. 1989. *Water administration in Australia: Agenda for change.* Occasional Paper No. 4, Centre for Water Policy Research, Armidale, University of New England.

New South Wales Parliament. 1885. *Royal Commission on the Conservation of Water.* First report of the commissioners. Sydney: New South Wales Government.

Paterson, J. 1987. *Law and water rights for improved water resource management.* Department of Water Resources Staff Paper 01/87. Melbourne: Department of Water Resources.

Powell, J. M. 1989. *Watering the garden state: Water land and community in Victoria.* Sydney: Allen and Unwin.

Powell, J. M. 1991. *Plains of promise, rivers of destiny: Water management and the development of Queensland, 1824–1990.* Brisbane: Boolarong Publications.

Randall, A. 1981. Property entitlements and pricing policies for a mature water economy. *Australian Journal of Agricultural Economics* 25: 195–220.

Reisner, M. 1993. *Cadillac desert: The American West and its disappearing water.* rev. ed. New York: Penguin.

Reeve, I., L. Frost, W. Musgrave, and R. Stayner. 2002. *Agricultural and natural resource management in the Murray-Darling Basin: A policy history and analysis.* Overview report to the Murray-Darling Basin Commission. Armidale: Institute for Rural Futures, University of New England.

Smith, D.I. 1998. *Water in Australia: Resources and management.* Melbourne: Oxford University Press.

Tyrrell, I. 1999. *True gardens of the gods: Californian-Australian environmental reform, 1860–1930.* Berkeley, CA: University of California Press.

Tisdell, J., J. Ward, and T. Grudzinski. 2000. *The development of water reform in Australia.* CRC for Catchment Hydrology Milestone Report. Canberra: CRC for Catchment Hydrology.

Ward, J. 2000. The evolution of water management in Australia. In *The development of water reform in Australia (CRC for Catchment Hydrology, Milestone Report)*, edited by J. Tisdell, J. Ward, and T. Grudzinski. Canberra: CRC for Catchment Hydrology.

Watson, A. 1996. Conceptual issues in the pricing of water for irrigation. In *Security and sustainability in a mature water economy: A global perspective*, edited by J. Pigram, 213–227. Armidale: Centre for Water Policy Research.

Watson, W. 1990. An overview of water sector issues and initiatives in Australia. In *Transferability of water entitlements: An international seminar*, edited by J. Pigram and B. Hooper, 213–227. Armidale, Centre for Water Policy Research.

Watson, W., and Rose, R. 1980. *Irrigation issues for the eighties: Focusing on efficiency and equity in the management of agricultural water supplies.* Paper presented to the Annual Conference of the Australian Agricultural Society, Adelaide.

Wilkinson, J. 1997. *Water for rural production in NSW: Grand design and changing realities.* Briefing Paper No. 26/97, Sydney, NSW: Parliamentary Library Service.

CHAPTER 4

The Legal Frameworks of Australian Water
Progression from Common Law Rights to Sustainable Shares

Jennifer McKay

THIS CHAPTER DESCRIBES AND ANALYZES Australian surface water and groundwater policies and laws, particularly in relation to agriculture. Building on the historical perspective provided by Musgrave in Chapter 3, I offer a legal perspective of the development phases in Australian water history. The chapter also goes beyond the period analyzed by Musgrave by providing an overview of the most recent policy initiatives. Particular attention is given to national water policy developments in 1994, 2004, and 2007, which are treated as separate policy episodes or eras.

The chapter is divided into four parts, reflecting four phases of water law. The two early phases coincide with Musgrave's detailed description of events up to the 1990s. Two additional phases are then delineated—one encompassing the period from 1994 to 2004 and another from 2004 onward. Consideration is given to the problems associated with implementing water laws and policies. The chapter also covers the legal distribution of power over water and attempts to describe, through case examples, aspects of the exercise of the powers that make water law and policy particularly complex in Australia.

Phase I—The Common Law

Upon colonization the common law of England was applied to groundwater and surface water by the high court of Australia. As noted earlier, this occurred in spite of the vast differences in climate and resulted in differing legal treatments for surface water and groundwater (see, for instance, *Gartner v. Kidman* 1962; *Dunn v. Collins* 1867). The surface water rule is known as the riparian doctrine and, notwithstanding the limitation noted by Musgrave in Chapter 3, it clearly imposed a reasonable user limit to preserve the river. By way of contrast, the groundwater rule failed to impose such limits, the owner of the land above having the right to use all available water. A groundwater rule of this type is generally considered as encouraging the tragedy of the commons. At common law, a landowner would not be able to win an action against a neighbor who ran the wells dry in a given region.

The authoritative case in Australia with respect to common law riparian rights is *Dunn v. Collins* (1867). Wearing J's judgment summarized the common law rights with respect to surface and groundwater. This decision gives an Australian application of the common law riparian doctrine as set down by Lord Kingsdown in *Minor v. Gilmour* (12 Moore PCC 133). For readers with a particular interest in this field a summary of other relevant Australian cases is provided in Table 4-1.

The common law specifically recognized that although the overlying landowner was not the owner of the water, he had an unlimited right to appropriate it (*Ballard v. Tomlinson* 1885) and use it for whatever purposes he pleased, either on or off the overlying land (*Chasemore v. Richards* 1859). This seemed to apply even if his sole purpose was malicious intent to injure his neighbors' wells (*Mayor of Bradford v. Pickles* 1895).

Phase 2—Vesting of the Use Power and Control Power in the States

In the second phase, the various states changed the systems to reflect broader water sharing privileges as irrigation was needed far beyond the river banks. This started from 1886, a full 100 years after white settlement.

Common law rules were replaced by a licensing and allocation system for both surface and groundwater. This occurred with vesting of the water in each state crown. During this phase, surface and groundwater regulation came under the control of the eight state and territory jurisdictions, each creating different laws and institutions to regulate water entirely within their domain. The complex negotiations and rivalry between jurisdictions that followed ensured that the laws were ostensibly introspective (Hallows and Thompson 1999) and pluralistic (Griffiths 1986).

Each state created its own legal and administrative systems for groundwater and surface water, generally without considering the conjunctive uses of the resource. Groundwater laws were usually passed only in response to a crisis (see Clark and Meyers 1969). Open access was the norm and states generally failed to cap the number of bores or the amount of water drawn. There was no payment for the water used, although farmers paid for the bore itself and associated reticulation infrastructure. The upshot was the continuation of the common law tradition established in the earlier phase, especially for groundwater, which generally remained under common law much longer than surface water. Each user was allocated water at will and individuals faced only weak incentives to moderate their use. In some places, such as the Great Artesian Basin, many uncapped bores were installed and water flowed freely day and night (Broughton 1999).

Table 4-1. Australian Surface and Groundwater Common Law Rules Relating to Water Use Prior to 1886

Common law legal rule Dunn v Collins (1867) 1 SALR 126	Surface water in a defined channel	Groundwater percolating through underground strata with no certain course and no defined limits	Groundwater in a defined channel	Groundwater in a defined channel connected to surface water in a defined channel
Use rights	Riparian doctrine applies. Riparian rights are, therefore, usufructuary rights that do not confer exclusive possession in the owner of the property through which the water is flowing. There is merely a right to use and benefit from the water and only to the extent that other riparian right owners are not detrimentally affected. Dunn v Collins (136)	Riparian doctrine does not apply. Hence unlimited use applies	Riparian doctrine does not apply. Hence unlimited use applies. Acton v Blundell (1843) A person who owns the surface may dig therein and apply all that is found to his own purpose at his free will and pleasure and, if in the exercise of such right, he intercepts or drains off the water collected from underground springs in his neighbors well, this inconvenience cannot become ground for an action. Quoted in: Dunn v Collins as per J. Wearing (140).	Riparian doctrine applies. Use is unlimited. Dickenson v. The Grand Junction Canal Company decided that: "If a person possesses a right to a stream jure naturae he has a right to its subterranean course. So that the test of the validity of a parties claim to the usufruct of water in a subterranean channel is whether he has or has not a legal right to the usufruct of a watercourse at the surface, which is fed by the underground channel." Dunn v Collins as per J. Wearing (141).
Ownership	Riparian is not an owner of the water. May use it for ordinary purposes such as domestic use. With domestic use the riparian may take all water. The riparian can also make extraordinary use of water but only if that does not affect those above or below the riparian user.	If the underground water has been accessed by a man-made well or bore then the water is accessed via an artificial structure. Thus, absolute rights will be vested in the proprietor who brought the water to the surface, as the riparian doctrine only applies to naturally occurring water.	If the underground water has been accessed by a man-made well or bore then the water is accessed via an artificial structure. Thus, absolute rights will be vested in the proprietor who brought the water to the surface, as the riparian doctrine only applies to naturally occurring water.	If the underground water has been accessed by a man-made well or bore then the water is accessed via an artificial structure. Thus, absolute rights will be vested in the proprietor who brought the water to the surface, as the riparian doctrine only applies to naturally occurring water.
Remedy between citizens for overuse by a fellow citizens	Yes; depending on the type of use.	No; unless trespass occurs to steal the water.	No; unless trespass occurs to steal the water.	No; unless trespass occurs to steal the water.

Source: McKay 2006. Presentation at the International Symposium of Groundwater Sustainability. Alicante, Madrid.

The grant or licensing system was created in each state for both surface and groundwater with the rights to the use, flow, and control of the water embodied in licenses. These licenses conferred entitlements on users. Importantly, entitlements were not property rights in a legal sense (Hohfeld 1919; Kelsen 1949; Dworkin 1975) as they were not at this point freely transferable; however, farmers generally viewed the license as a right, and in subsequent eras when allocations were amended, there was much community angst about the perceived loss of these rights.

The surface and groundwater laws were usually administered by multiple government agencies with the resulting systems being very complex for the landowner. For instance, in Victoria the Groundwater Act of 1969 was described as " . . . manifesting an unhappy and complex division of functions between the Department of Mines and the State Rivers and Water Supply Commission which sets a sophisticated barrier of official correspondence between farmer and the water under his feet" (Clark and Meyers 1969). The Commission was nominally responsible for investigating groundwater resources but, by administrative arrangements, the Department of Mines took control of this matter.

Notwithstanding the complexities within jurisdictions and the inconsistencies between authorities, each state resolutely retained control over its water resources during this phase. The states decided to federate in 1901, but the issue of water was a key stumbling block. The powers of the new federal government were listed in Section 51 and these were broad in that they covered thirty-nine areas including trade and commerce. The powers of such an entity could be construed widely (Crawford 1991) and it was because of the wide coverage of Section 51 that the states insisted on Section 100 being inserted as follows:

> The Commonwealth shall not, by any law or regulation of trade or commerce, abridge the rights of the State or of the residents therein to the reasonable use of waters of rivers from conservation or irrigation.

The parliamentary debate circumscribing this matter reveals that Section 100 was inserted because New South Wales, Victoria, and South Australia feared that Commonwealth Laws, under Section 51, might affect their common interest in water for irrigation (Lane 1986). The contest revolved around the Commonwealth's power over water for navigation and the state's desire to use water for irrigation. The object of Section 100 is thus twofold: First, it was designed to limit Section 51, and second, it was intended to limit the inherent paramountcy of the Commonwealth navigation power (Quick and Garran 1976).

Several cross border agreements emerged throughout this phase, with some remaining in place today. For example, South Australia and Victoria forged an agreement with regard to a vital aquifer in the southeast, and both states joined an agreement with New South Wales, Queensland, and the Northern Territory over the Great Artesian Basin. As already noted, the Murray-Darling Basin agreements were also forged during this phase.

Despite Section 100, the Commonwealth has intervened in state water management through Section 96 of the Constitution which gives the Commonwealth power to grant financial assistance to the states and impose conditions. In this context, the emergence of salinity problems was identified as a key issue, especially along the River Murray, and in 1978 the Commonwealth passed the National Water Resources Financial Assistance Act. The purpose of the Act was to fund a broad range of works aimed at conserving water and mitigating salinity and floods, particularly in the Murray-Darling Basin.

The operation of the Murray-Darling Basin Agreement and the practical role played by the Commonwealth are illustrative of its policy influence in water. The Agreement covers all natural resource management and, among other goals, aims to reduce the salinity impacts of river water use for irrigation.

In response to an audit of the Murray-Darling Basin, an interim Cap on extractions was imposed in June 1995. The Cap limited the amount of water that could be diverted for consumptive uses to the quantum diverted on June 30, 1994 (MDBC 1998). The Cap aims to restrain further increases in water diversions, but it does not constrain new developments, provided the water for them is obtained by using water more efficiently or by purchasing water from existing developments.

Following an independent review of equity issues that attend the Cap, it was made permanent for New South Wales, Victoria, and South Australia from July 1, 1997. This has been heralded as one of the most important decisions ever made in Australian water history (see, for instance, Blackmore 2000).

Under the Cap arrangements, each state is required to monitor and report to the Murray-Darling Basin Commission (MDBC) on diversions, water entitlements announced, allocations, trading of water within, to, and from the state, and detail compliance with the target. The state must also report on measures undertaken, or proposed, to ensure that the water taken does not exceed the annual diversion target for every ensuing year. The MDBC appoints an Independent Audit Group (MDBC IAG) which annually audits the performance of each state government. There is also power to order a special audit if:

- the diversion for water supply to metropolitan Adelaide has exceeded 650GL, or,
- the cumulative debit recorded in the Register exceeds 20 percent of the annual average for a particular river.

Many special audits have been conducted on the basis of the latter.

In 1999 the MDBC commissioned a review of the operations of the Cap and drew three main conclusions. First, the Cap had supported the Ministerial Council's aim of achieving the ecological sustainability of the Basin's river systems. Second, although the Cap does not necessarily provide for a sustainable basin ecosystem, its implementation had been the essential first step in achieving this goal. Third, without the Cap there would have been a significantly increased risk that the environmental degradation of rivers systems would have been worse (MDBC 1999).

There are various legal dimensions to the Cap. One of the most pressing is that the Cap requires state-level implementation and yet the Commonwealth has no power to mandate that the states resource the necessary compliance work.

In this context it is interesting to note the IAG reports, which provide a vehicle for each state to voice concerns about the actions of others. A key concern expressed by South Australia has been that New South Wales has exceeded its Cap in a number of rivers; notably, the Border Rivers, Gwydir, Lachlan, and the Barwon-Darling. Moreover, it was observed in the IAG reports that New South Wales had not finalized its Cap compliance through an independent verification process by 2000–2001 and South Australia remained concerned about "New South Wales' ability to bring diversions back into balance with Cap limits into the future" (MDBC IAG 2002).

In 2000/2001 the IAG paid special attention to activities in New South Wales, conducting Special Audits of the Namoi, Lachlan, Barwon-Darling, and lower Darling Rivers. The IAG established that the long-term diversion Caps had been exceeded in these valleys, however, the New South Wales government broadly argued that any violation of the Cap had been caused by poor quality data and the state reportedly assigned additional resources over the 2001/2003 period to resolve these problems. Interestingly, although the IAG recommended a major commitment of resources in each jurisdiction to enhance data quality, New South Wales only agreed to continual improvements in data collection. The most recent report (MDBC IAG 2005/06) echoes these concerns and prompted the prime minster to call for a federal takeover of the power of the states in January 2007.

Great Artesian Basin Cooperative Agreement

Emerging problems with groundwater also prompted individual state governments to cooperate to undo some of the legacy of the open access uncapped bores developed during Phase 2. Four state governments created a scheme to Cap bores in the Great Artesian Basin (GAB) and hence restore pressure. The area was mined by more than 4,000 flowing bores. By 1990, 1,000 of these had stopped flowing. The major use of the water is for livestock as the dissolved minerals make it too salty for other purposes. The main concern in the Artesian Basin had been the loss of pressure, which is itself dependent on the drawing of water from nearby bores as the lateral movement of water is very slow.

The Great Artesian Basin Rehabilitation Program started in 1989 and aims to encourage the capping of bores and piping of water. The program is funded by the Commonwealth and states. The cost sharing scheme for bore capping and pipes involves an 80 percent contribution by governments and a 20 percent investment by farmers (in New South Wales and Queensland). In South Australia, the capping program operates in only one area, with the state and Commonwealth providing all funding. The take-up of the program has been slow: Of the 1,380 uncontrolled bores operating in 1989 only 250 were repaired by 1997. Subsequently, the Great Artesian Basin Consultative Council was established to manage the GAB and consists of groundwater users, industry, local government, tradition owners, conservation groups, and governments. Funding for this entity is shared equally between New South Wales, Queensland, and South Australia.

In sum, Phase 2 was characterized by a move to vest water in the state governments although the enthusiasm for this approach varied between states and occurred more rapidly with surface water than with groundwater. The phase was also punctuated by increased interstate cooperation and an expanded role for the Commonwealth; an approach which became more overt in the 1990s.

Phase 3—Council of Australian Governments (CoAG) Reforms 1994–2004

The Commonwealth government's stake in Australia's water affairs changed significantly with the incorporation of water management into the CoAG competition framework. Arguably, this represents a definitive shift in Australian water policy and formalizes the earlier encroachment of the Commonwealth into water affairs. National competition policy, which included the part or full sale of several public enterprises, created a pool of funds by which each state could be encouraged to follow national water protocols. The CoAG reforms of 1994 were thus the primary motivation to amend policy and resulted in the restructuring of water management regimes and institutions through new water laws in each state. The reforms insisted that each state ensure that future water projects were based on Environmentally Sustainable Development (ESD) principles, in conjunction with much more private sector participation and community involvement in water management and planning at a regional level.

There were eight core elements to the first CoAG framework. First, pricing reforms were introduced with the ambition of full cost recovery and the removal of cross-subsidies, or their codification in a transparent manner. Second, each jurisdiction was to adopt a comprehensive system of water allocations, including allocating water for environmental purposes. Third, the nexus between land and water titles was to be broken so that water entitlements could be traded. Fourth, it was expected that water service provisions would be structurally separated from resource management, standard setting, and regulation by 1998. Fifth, two part tariffs were to be introduced for urban water users wherever cost effective. Sixth, each state was required to put in place arrangements for water trade. Seventh, rural water charges were to be progressively amended to achieve cost recovery by 2001. Finally, all future irrigation investments were to satisfy economic and environmental sustainability criteria before proceeding.

Understanding of ESD

Amongst the challenges of implementing the early CoAG reforms were problems related to the scope of key terms, like ESD and the vast number of licenses relating to water (Shi 2005). The fundamental premise of ESD is that economic development must be balanced against the protection of biological diversity, the promotion of equity within and between generations, and the maintenance of essential ecological processes (Brundtland 1987). The Commonwealth government working groups on ESD drafted seven principles to guide ESD in 1992 and the acceptance of these principles by CoAG reflects a move away from focusing exclusively on economic efficiency; rather, there is a need to achieve a balance between social, economic, and environmental needs (Harris and Throsby 1997). These principles have accordingly guided the collective thinking of governments in the formulation of water policy throughout this and the subsequent phase.

Responses to the Changes in Water Law and Policy

From a legal perspective there are two key cases dealing with the community's response to this policy shift in Victoria and New South Wales.

In the first case, *Ashworth v. State of Victoria* (2003), a farmer challenged the validity of the new laws in Victoria, which vested the rights to water captured in farm dams to the state. The plaintiff had four dams on his property and used them to capture water for irrigating pasture and crops, a practice he had undertaken for the past thirty years. The dams were filled by runoff from rain falling on the land and by water seeping from springs.

Amendments to the Water Act Victoria, however, restricted the previous rights of farmers to capture water in dams. This Act has been described as the most comprehensive in the nation insomuch as it aimed to set up a framework and management regime to ensure that upstream uses do not affect downstream users in Victoria (Pisaniello and McKay 2005). The government of Victoria passed the Water (Irrigation Farm Dams) Act in 2002. The purpose of the Farm Dams Amendment Act (Explanatory memorandum 2002) was remedial, to complete Victoria's water allocation framework by requiring the licensing of all irrigation and commercial use within the catchments. Existing irrigation and commercial dam owners who were not currently required to be licensed had a choice of applying for either:

- a registration license, or
- a standard take and use license, which is transferable.

The plaintiff, however, did not obtain a license and objected to the interference with what he saw as his right to use the water in his dams as he saw fit. He sought a declaration from the court. The case highlighted the role of water law in Victoria and its capacity to increasingly control and conserve water in the interests of the public. The Act was upheld and the plaintiff was required to gain a license and pay costs.

In a similar vein, a case occurred in New South Wales under the new Water Management Act. This Act had the new objective of conserving water resources and making them available at a level that would ensure sustainability for future generations.

This second case was pursued and prosecuted diligently on behalf of Murray Irrigation Limited (MIL), a large private irrigation company which diverts water from the River Murray and allocates it to 2,400 farmers. Each farmer is a shareholder of MIL and is entitled to a water allocation depending on his/her shareholding.[1]

MIL had responded to protracted drought conditions by implementing an 8 percent reduction in the amount of water distributed to shareholders. The argument in the case revolved around the unlawful taking of water and the defendants—Meares Pty Ltd—attempted to argue that they could not be held liable for the actions of their employee, who took water in excess of the amount licensed by tampering with a Dethridge wheel.

Under the new Water Management Act, the offences were criminal and of a strict liability nature. Strict liability is a severe test and it is notable that the Act applies this approach. Strict liability means that there is no requirement for the prosecution to prove mental or fault element in the form of intention, recklessness or negligence (*He Kaw The v. R* [1985]) Commonwealth Law Reports 523). The case found the defendant liable. Moreover, the case supports the view that any water allocation is a mere license able to be reduced by a relevant authority under powers designed to promote ESD.

In sum, the outcome of these cases and the earlier discussion show that Phase 3 formalised the enhanced status of the Commonwealth in water policy and law formulation via the CoAG framework and simultaneously raised the legal standing ascribed to ESD principles and encouraged regional delivery.

Phase 4—The National Water Initiative and Commonwealth Water Bill 2007

Notwithstanding the move toward a more uniform approach to water law and water resource management promised by the early CoAG framework, complexity and lack of coherence between jurisdictions persisted. Consequently, CoAG announced a second major series of reforms, known as the National Water Initiative (NWI), in 2004. The NWI is also a product of the CoAG process and aims to achieve national compatibility in the markets, regulatory, and planning schemes to achieve sustainable management of surface and groundwater. The NWI Agreement was signed by all governments, with the exceptions of Tasmania and Western Australia, at the June 29, 2004 CoAG meeting. These two states subsequently became signatories to the NWI in 2005 and 2006, respectively.

The NWI specifies that consumptive use of water requires a water access entitlement to be described in legislation as a perpetual share of the consumptive pool of a water resource (NWI paragraph 28). The NWI represents a shared commitment by the Australian commonwealth government and state/territory governments to ESD principles insomuch as it recognizes:

- the continuing national imperative to increase the productivity and efficiency of Australia's water use;
- the need to service rural and urban communities; and
- the importance of ensuring the health of river and groundwater systems, including the establishment of clear pathways to return all systems to environmentally sustainable levels of extraction (paragraph 5, NWI).

The NWI is a comprehensive reform agreement containing objectives, outcomes, and agreed actions to be undertaken by governments across eight interrelated elements of water management. Some of the core elements of the NWI and their legal dimensions are briefly discussed here.

Clear Specification of Water Access Entitlements

Separation of land title from water title has been pursued by state and territory governments since the 1994 CoAG water reform framework. The NWI further specifies that consumptive use of water requires a water access entitlement. Moreover, it is described in legislation as a perpetual share of the consumptive pool of a water resource (NWI paragraph 28). A review of progress on this front reveals mixed success across states. The NWI also specifies the characteristics that water access entitlements should have (NWI paragraph 31), including that they: be exclusive; are able to be traded; are able to be subdivided or amalgamated; are able to be mortgaged to access finance; and are recorded in public water registers.

In most states and territories, the conversion of existing water entitlements into share-based entitlements, as required under the NWI, is still a work in progress.[2] For example, in Queensland and New South Wales, conversion of entitlements is occurring only when water plans are completed for catchments and groundwater management areas—these water plans establish the available consumptive pool of the water resource. In South Australia, the process of deciding the consumptive use of water commenced in 2005. The task also involves an assessment of long term environmental requirements. All major resource users have been allocated a volumetric allocation, including the environment, and all significant water resources are being transformed into volumetric allocations for all purposes (SA Government 2005).

The NWI also requires that water provided to meet environmental and other public benefits must have statutory recognition, and have at least the same degree of security as water access entitlements for consumptive use (NWI paragraph 35). This is to ensure that water for environmental outcomes is not made less secure in the wake of greater security for consumptive water entitlements.

Efficient Water Markets

As it stands, there are a range of institutional barriers to the trade of permanent water entitlements out of many irrigation districts in Australia—either in the form of trading rules, policies governing public irrigation authorities, or policies contained in the memoranda and articles of association of some private irrigation corporations (notably in New South Wales; Shi 2005). As signatories of the NWI, the state governments of New South Wales, Victoria, and South Australia are taking steps to free up trade out of their irrigation areas.

Initially, trade out of each irrigation area is intended to be limited to no more than 4 percent of each area's total water entitlement. This measured step is provided in the NWI in order to help manage concerns about the adjustment of regions to trade, and to enable monitoring of the socioeconomic impacts of trade.

Expansion of water trade will also rely heavily on reducing the transaction costs of trades. In particular, the NWI requires compatible water registers between states and other compatible institutional arrangements in order to enhance trading opportunities.

Water Accounting

Along with secure property rights, most markets require an agreed metric. Most states are currently in the process of expanding metering of water used for irrigation. Australia has almost universal metering of water used in residential and business settings in major metropolitan areas. Adequate metering practices and accounting systems for water are, of course, necessary for effective charging, and to support water trading (e.g., to ensure that water that is traded is available to be traded, is delivered to the buyer, and that information about water trades is made available to inform the market).

Less sophisticated measurement and monitoring of water may be entirely appropriate in catchments where the resource is relatively undeveloped and there are few production pressures. In such cases, the need to improve monitoring is driven by the need to better understand the resource so as to better manage its environmental values.

Clearly, on the basis of the evidence presented by Letcher and Powell in Chapter 2, improvements in this field are an imperative in many catchments.

Clear Assignment of Risk for Changes in Water Allocation

As noted earlier, the creation of share-based water access entitlements establishes a secure right to access the water resource. In the NWI, governments have also committed to establish a level of security around the size of the consumptive pool of water that entitlement holders can access. To this end, the NWI establishes a framework for assigning the risks of future reductions in the availability of water for consumptive use (NWI paragraphs 46–51). The difficulties associated with implementing this element of the NWI are addressed in detail by Quiggin in the following chapter.

Improved Water Pricing Policies

There have been significant improvements in water pricing arrangements since the 1994 CoAG water reform framework. In the NWI, governments have committed to continue with pricing reform, in particular:

- to continue movement to pricing that recovers the full costs of water storage and delivery for rural and regional systems;
- to continue movement to pricing that achieves a commercial return on assets (while avoiding monopoly rents) for metropolitan, rural, and regional water storage and delivery;
- pricing that recovers a proportion of the costs of water resource management and planning—cost recovery for such activities to manage the consequences of commercial water extraction has become a legitimate proxy for more direct externality pricing in rural areas;
- nationally consistent benchmark reporting on the service quality and pricing of all water service providers; and
- moving toward more nationally consistent approaches to pricing across all these areas (Thompson 2005).

Water pricing reform is currently a very active area for most state and territory governments. The overall intent is to ensure that prices set by mechanisms other than the market (i.e., by governments, public/private water service providers, and/or independent pricing bodies) do not lead to perverse outcomes either in secondary water markets, or for water-related investment activity.

Just under half of the NWI's 70 or so actions involve national actions or other action by governments working together. This reflects an emphasis on greater national compatibility in the way Australia measures, plans for, prices, and trades water. It also represents a greater level of cooperation between governments to achieve these ends. This process will be driven by the new National Water Commission and the $2 billion to be invested over six years through the Australian Water Fund. The National Water Commission is established under Commonwealth government legislation (i.e., the National Water Commission Act 2004). It is an independent statutory authority reporting to the prime minister and, in some instances to CoAG, via the prime minister.

The Legal Implications of the NWI Requirements

This fourth phase of water law in Australia is innovative as it endeavors to create a perpetual water right in the hands of each water user; however, this perpetual right pertains only to an annually determined share of the consumptive pool, and in that sense it is not a right to a volume at all. There remains some conjecture about how a court would view these arrangements. In the aforementioned cases of *Ashworth* and *Meares* there is a strict adherence to the clear intent to regulate water use. However, fundamental questions remain about the definition and fulfillment of these intentions under the NWI.

The process in Australia requires interpretation by courts of the relevant water statutes, statutory interpretation, and regulations in each jurisdiction. These all differ (because of Section 100) and progress toward a common lexicon is ongoing. Each decision is a legal formant (Watson 1995) and there will be eight separate formulations operating in Australia (Griffith 1986; Watson 1995). The cases described earlier in this chapter also provide evidence of some judicial discomfort with the legal definition of sustainability and variations in the definition between jurisdictions.

In addition to the difference in relative weights of legal formats, consideration must be given to the different water governance structures in each state and the history of these institutions. The relationship between the institutions and their governing laws is of crucial relevance to the interpretation of the law. It is characteristic of all legal systems that the governance systems are hard to verbalize, difficult to quantify but, patently, of enormous importance in a legal context (Hart 1961; Singer 1982; Sacco 1991).

There seems little doubt that administrative law will be used to test the boundaries of the water planning processes and, in particular, the meaning of the perpetual share; however, administrative law can only review the legality of decisions, not their merits. As detailed by Brennan (*Attorney General v. Quinn* 1990, *35–36*):

The duty and jurisdiction of the Court to review administrative action do not go beyond the declaration and enforcing of the law which determines the limits and governs the exercise of the repository's power. If in doing so, the Court avoids administrative injustice or error, so be it; but the Court has no jurisdiction simply to cure administrative justice or error. The merits of the administrative action, to the extent that they can be distinguished from the legality, are for the repository of the relevant power and, subject to political power.

There is one recent example of the use of administrative law to challenge elements of the reform process. This matter related to the water planning process which, under the NWI and earlier CoAG agreement, requires consideration of the long-term environmental sustainability of the resource.

In New South Wales, the case of *Murrumbidgee Groundwater Preservation Association v. Minister for Natural Resources* revolved around the validity of a water plan. The Murrumbidgee Association challenged the plan on many grounds, primarily by using administrative law. Administrative law deals with procedural fairness and the grounds alleged were:

- Extraneous purpose of the minister in making the plan that was to avoid the community drafted plan;
- The formula for reserving waters for the environment contained a mathematical impossibility;
- Uncertainty of timing of operation of the plan; and
- The imposition of uniform reductions in water allocation was irrational.

The area involved was within the power of the minister to draft a groundwater plan himself. The plan addressed sustainable management of groundwater and identified limits on extraction. The overall aim of the plan was to reduce actual use over 10 years to the annual average recharge, less a quantity preserved for environmental benefit. Groundwater users were subject to pro rata reductions of entitlements over a ten-year period. By year nine of the plan, all users were to be entitled to only 52 percent of their original entitlements. The plan also incorporated adjustment mechanisms, such as the creation of a market in access licenses and supplementary water access licenses.

All of the grounds for appeal were dismissed. The grounds were dealt with as follows:

- Extraneous purpose. The Appellant alleged that the minister made the plan to avoid the notification, public exhibition, and considerations as required under a plan made by a management committee. It was held that the power to establish a management committee to draft a plan is discretionary and a plan formulated by the minister is valid.
- The literal construction of the clause provided an absurd result, so the court applied a purposive construction.
- The timing was considered to be capable of being certain and so valid.
- The case here was that it was irrational to treat the groundwater source as a single body of water, as aquifer recharge was site specific and that an activity in

one area will result in changed conditions elsewhere. Historically, the groundwater system was managed in zones because in some areas use of entitlements would be unsustainable. The argument applied the precautionary principle to protect the resource in the absence of scientific data.

• The single system was also argued to be irrational, as it was not based on water availability; however, the court upheld the pro rata reduction on the grounds that the court has a confined role and it was for the minister to balance the desired environmental outcome and the chosen method of achieving it.

Notwithstanding the outcomes of the case mentioned earlier, which was pursued under administrative law, the interplay between the statements in the various water acts and the general law principles of natural justice is likely to be an area of much litigation in the future. The upshot of the NWI is that any rural water user has a perpetual percentage share of water in a defined resource and the quantum of water will be determined each year to account for the environment and then other users. Numerous elements within these processes have yet to be fully tested from a legal perspective.

Commonwealth Power Over the Murray-Darling Basin

On 25th January 2007, the prime minister announced a national plan for water security. This was based on the assumption that states would refer powers to overcome limitations resulting from Section 100 (www.pm.gov.au/national_plan_water_security.pdf). The prime minister expressed exasperation with the current issues (see earlier discussion), especially the parochial nature of water decisionmaking. The plan offered $10 billion to be spent on water saving infrastructure, water monitoring, purchase of entitlements and reform of the decisionmaking processes. Three states agreed readily but Victoria refused to sign and so the prime minister revised the bill to rely on existing federal constitutional powers. At this time submissions on the bill are not available.

Conclusion

The evolution of water law in Australia can be broadly conceptualized as comprising four distinct phases. The first phase assigned rights to water users and gave very little consideration to the environment. The riparian rule created a right and duty to not sensibly diminish, although the groundwater rule encouraged unfettered use. The second phase was the period of vesting water in allocation systems to widen its use. This approach created licenses representing a mere claim on the water in law rather than a right per se. In the minds of many farmers, however, they had a right to the water and it was rarely taken away or challenged. During this phase groundwater was also subject to licenses, although this was, in practice, little different than the open access regime of the earlier phase. Rarely was there a charge imposed on water use and the rationale that underpinned this phase largely encouraged the creation of over-allocated systems, especially in New South Wales.

The third phase was crystallized under CoAG in 1994 and imposed an obligation to include ESD in water allocation decisions as well as other market reforms to promote private sector provision of water and water trading. This was the time of community agitation over drought and blue-green algae. Stressed rivers were identified in many states and plans made to alter allocation processes. There was considerable angst over loss of consumptive water rights and acrimonious debates about compensation. Importantly, it signaled the formal enhancement of the role of the Commonwealth in water policy formulation.

The fourth phase imposes a clear rule for the allocation of water for consumptive use. This rule is to be consistent across state jurisdictions and amounts to a perpetual share of the consumptive pool of the resource. Arguably, this represents the weakest legal right for users to date. The NWI approach requires conformance to a local plan (drafted under various state laws) which allocates the quantum of water in a local resource over a set time period: however, as a product of Section 100 of the constitution and attendant political issues, variations in the definition of ESD between states has the potential to prove problematic.

The direct outcome for the farmer is that previous allocated volumes of water are replaced by a percentage share of an annual sustainable yield of surface water or groundwater. Accordingly, irrigators (and other right holders) have a right to that share but the quantum varies annually. The annual quantum in each river system is decided by a process of water plans, which endeavor to integrate scientific data with economic and social information as required to achieve ESD. These processes continue to be developed and have already been contested in New South Wales and Victoria. The processes themselves also differ markedly between the states.

This final phase also confirms the expanded role of the Commonwealth government, despite Section 100, and attempts to have the state's power referred. This has manifested in considerable persuasion over the shaping of early CoAG reforms and the more recent NWI. Although this influence can undoubtedly be traced to the vertical fiscal imbalance that typifies Commonwealth–state relations, the future legal standing of these arrangements has yet to be fully tested.

Acknowledgments

Adam Gray, Fiona Partington and Michael Griffin provided research assistance for sections of this chapter. Thanks also to support from the CRC Irrigation Futures, Kent Martin (SAFF), the Natural Resource Management Council of SA, the University of South Australia, the ARC and Land and Water Australia.

Notes

1. For a more comprehensive description of the operational aspects of MIL see McLeod and Warne in Chapter 7.

2. For a more complete description of the role and organisational form of natural resource management organisations involved in establishing water entitlements see, for example, McKay (2006); McKay (2007).

References

Australian Government. 2006. NWI Implementation Plan. http://www.nwc.gov.au (accessed November 25, 2006).

Blackmore, D. 2000. Managing for scarcity—Water resources in the Murray-Darling Basin. *CEDA Bulletin* July: 50–54.

Broughton, W. ed. 1999. A Century of Water Resources Development in Australia 1900–1999. The Institution of Engineers Australia.

Brundtland Report. 1987. *Our common future.* World Commission on Environment and Development (WCED).

Clark S. D. and A. J. Meyers 1969. Vesting and divesting: The Victorian Groundwater Act 1969, 7. *Melbourne University Law Review* 238.

Crawford, J. 1991. The constitution and the environment. *Sydney Law Review* 13:10–25.

Dworkin, R. M. 1975. Hard Cases 88 *Harvard Law Review* 1057; reprinted in his Taking Rights Seriously, supra, 81.

Explanatory Memorandum to the Water (Irrigation Farm Dams) Act. 2002. Amending the Water Act. Victorian Government.

Griffiths J. 1986. What is legal pluralism? *Journal of Legal Pluralism* 24:1–4.

Hallows P., and D. Thompson. 1999. The history of irrigation in Australia, Australian National Committee on Irrigation and Drainage, Mildura.

Harris, S., and D. Throsby. 1997. The ESD Process: Background, implementation and aftermath. In C. Hamilton and D. Throsby (eds.), The ESD process: Evaluating a policy experiment. Workshop organised by the Academy of the Social Sciences and the Graduate Program in Public policy, Australian National University, Canberra, 28-29 October 1997.

Hart, HLA. 1961. *The concept of law.* Oxford: Oxford University Press.

Hohfeld, Wesley Newcombe. 1919. *Fundamental legal conceptions as applied in kudicial reasoning,* ed. W. W. Cooke, New Haven: Yale University Press.

Kelsen H., 1949. *General theory of law and state.* trans. Anders Wedberg. Cambridge, MA: Harvard University Press.

Lane, P. H. 1986. *Lane's commentary on the Australian constitution.* North Ryde NSW: Law Book Company.

McKay J. M. 2006. Issues for CEOs of water utilities with the implementation of Australian water laws. *Journal of Contemporary Water Research and Education* 135:120-137.

McKay J. M. 2007. Water governance regimes in Australia implementing the National Water Initiative. *Water* 150-157.

MDBC (Murray-Darling Basin Commission). 1998. *Managing the resources of the Murray-Darling Basin.* Canberra: Murray-Darling Basin Commission.

MDBC IAG (Murray-Darling Basin Commission Independent Audit Group). 2002. Reports on CAP Implementation. http://www.mdbc.gov.au/nrm (accessed 25th November, 2006).

MDBC IAG (Murray-Darling Basin Commission Independent Audit Group). 2005-06. Report on CAP Implementation 2005-2006. http://www.mdbc.gov.au/__data/page/1658/IAG2005-6-full.pdf

Pisaniello J., and J. M. McKay. 2005. Australian community responses to upgraded farm dam laws and cost effective spillway modelling. *Water Resources Research* 12(2): 325–340.

Quick J. and R. R. Garran. 1901 (Reprinted 1976). The annotated constitution of the Australian Commonwealth. Sydney: Legal Books.

Sacco R. 1991. Legal Formants; A dynamic approach to comparative law. *American Journal of Comparative Law* 39:1–34.

Shi, T. 2005. Simplifying complex water entitlements. *Water* 32(5): 39–41.

Singer, J. W. 1982. The legal rights debate in analytical jurisprudence from Bentham to Hohfeld. *Wisconsin Law Review* 975–1059.

Thompson M. 2005 Sustainability, markets and policies. Paper presented at OECD Workshop on Agricultural and Water; Adelaide, www.oecd.org/agr/env and National Water Commission www.nwc.gov.au

Victoria's NWI Implementation Plan. 2006. Our water our future—A Victorian government initiative. www.nwc.gov.au (accessed November 25, 2006).

Cases

Acton v. Blundell (12 M&W; 324)

Ashworth v. Victoria (2003) VSC 194

Attorney General v. Quinn (1990) 170 CLR at 30

Ballard v. Tomlinson (1885) LR 29 Ch.D 115

Chasemore v. Richards (1859) 7 H.L.C 349

Commonwealth v. Tasmania (1983) 46 ALR 625

Dickenson v. Grand Junction Canal Company (7 Exchqu, 282)

(*Strachan v. Minister*) or require better information (*Elandes and Seidel v. Minister*)

Dunn v. Collins (1867) 1 SALR 126

Gartner v. Kidman (1962) 108 CLR 12

Griffith v. Civil Aviation Authority (1996) 41 ALD 50

He Kaw Te v. R (1985) Commonwealth Law reports 523

Mayor of Bradford v. Pickles (1895) A.C 587, 594

Minor v. Gilmour 12 Moore PCC 133

Murrumbidgee Groundwater Preservation Association Inc v. Minister for Natural Resources (2005) NSWCA 10

Neibieski Zamek Pty Ltd v. Southern Rural Water (2001) Victorian Civil and Administrative Tribunal 31 May

Rashleigh v. Environment Protection Authority (2005) ACTSC 18

Tasmanian Trust Case 2003

CHAPTER 5

Uncertainty, Risk, and Water Management in Australia

John Quiggin

T HE WEATHER IS INHERENTLY UNCERTAIN, and, as observed by Fletcher and Powell in Chapter 2, few aspects of weather are more uncertain than rainfall in Australia. Of the world's major river systems, the Murray-Darling, has not only the lowest, but by far the most variable rainfall in its catchment. It is natural, then, to expect that the management of uncertainty should be a central issue in Australian water management, and, in important respects, this is the case.

Paradoxical outcomes, however, are never far away in relation to water. The oft repeated fact that Australia is the driest continent in the world has naturally led governments to take responsibility for water supply, which has then encouraged the view that, as a necessity, water should be supplied free of charge, even for nonessential or frivolous purposes: the exact opposite reasoning that might be expected in a country where water is particularly scarce.

Similarly, the very prevalence of uncertainty has, in some cases, led to demands for perfect security, through technological approaches to drought proofing, regulatory policies designed to protect water users from risk and uncertainty, or market-based policies based on the assumption that clearly defined property rights can reduce uncertainty to bundles of state-contingent claims. The quest to eliminate uncertainty is futile. Uncertainty can be managed, allocated, and sometimes mitigated; it can never be eliminated.

In recent years, the central role of risk and uncertainty in water management has been recognized more explicitly. Principles of risk allocation and mitigation have played a major part in the formulation of the National Water Initiative (CoAG 2004), details of which are summarized by McKay in the previous chapter. This chapter endeavors to shed light on the problem of uncertainty in Australian water management and the attendant policy responses.

The main focus of attention is on the Murray-Darling Basin, although the policy lessons are applicable in many other Australian contexts. The scope of the Basin was described earlier and the problems of the Basin have been a major concern for Australian governments, helping to shape responses to broader policy issues relating to water management.

Sources of Uncertainty

Economists frequently distinguish between risk and uncertainty, a distinction first drawn by Knight (1921). The term *risk* refers to the special case where uncertainty about future events may be described in terms of a known probability distribution over a set of possible outcomes, normally derived from previous experience. The term *uncertainty* covers more general cases where probabilities may be unknown, or the subject of disagreement, or where the set of possible outcomes is not exhaustive, leaving open the possibility of some unforeseen event that may change the nature of the problem.

Uncertainty is an inherent feature of water management in Australia. Both medium-term cycles, such as the Southern Oscillation, and longer term climate change contribute to uncertainty, in addition to that arising from seasonal fluctuations. Agricultural producers are also subject to the uncertainty of demand, and the policy process itself generates uncertainty.

Annual and Seasonal Variation in Rainfall

The Murray-Darling system has not only the lowest but the most variable and unpredictable rainfall of any of the world's major river systems. Rainfall is variable and unpredictable on a range of time scales, from the very short term in the form of annual variations, to multiyear cycles such as those arising from the El Nino/Southern Oscillation phenomenon. The distribution of natural flows in the system is characterized by high variance and high skewness. The mean annual outflow under natural conditions is about 13,000GL, and the median is about 11,000GL. From a Basin-wide perspective, there has been a shift from a steady variation in mid-range flows under natural conditions to a dominance of low flows and occasional high flow events (Thomson 1994, *10*).

In its natural state, the Murray-Darling system typically displayed marked seasonal variability in flows, with high flows in spring, following the melting of snow in the Australian Alps, and low flows in late summer and autumn. However, this pattern was commonly disturbed by droughts and floods, which were, and remain, unpredictable. Not only has irrigation reduced unpredictable seasonal variation, it has also altered the dominant pattern of flow, reducing the spring peak and increasing flows in summer, when water is in greater demand for irrigation and urban water use.

Climate Change

Until recently, it seemed reasonable to assume that, despite marked variations over timescales up to a decade, climate was reasonably stable in the long run.

However, there is increasingly clear evidence that human activity, including the burning of carbon-based fuels and the clearance of forests, has led to an increase in the atmospheric concentration of greenhouse gases such as carbon dioxide, resulting in anthropogenic climate change, commonly referred to as global warming (IPCC 2001).

Although it is generally agreed that anthropogenic climate change is taking place, there is little certainty concerning the current and likely future rate of anthropogenic global warming. Even greater uncertainty surrounds the effects of climate change, natural and anthropogenic, on rainfall patterns, particularly at the regional level.

Jones et al. (2001) project a decline in winter and spring rainfall in the Murray-Darling Basin by the year 2030, but remain uncertain about impacts in summer and autumn. Average temperature, and therefore evaporation, is expected to increase in all areas of Australia. The two effects interact, with greater increases in evaporation likely in regions and seasons with less rainfall.

Demand Uncertainty

There is also uncertainty about the demand for water. Agricultural demand for water is a derived demand, ultimately determined by the demand for the products of irrigated agriculture and by the opportunity cost of the best dry land alternative. Given a derived demand curve, and a market in which water rights are freely traded, the quantity of water demanded by any individual producer will be determined by the market price.

Because of the variable and state-contingent nature of water supply in Australia, water cannot be treated as a homogeneous commodity. Water rights may have different levels of security, from season to season and over time. Depending on the purpose for which water is used, different sets of water rights may be appropriate (Freebairn and Quiggin 2006).

Demand for water for urban use depends, to a large extent, on policy decisions. At present, government policy in Victoria prohibits the diversion of water from the Murray Basin to Melbourne, even though the value of water in the urban market (net of pumping, storage, treatment, and reticulation costs) is considerably greater than its value in agricultural use.[1] Similarly, although Adelaide draws on the River Murray for some of its water supply, there is a considerable gap between the marginal value of water in urban and rural uses.

A change in policy, allowing urban and rural water uses to compete directly in the market, would lead to a significant upward movement in the demand for water, at least in areas where the cost of moving water to urban uses is low. In the policy environment that has existed until now, strong opposition from irrigators has rendered any substantial changes in policy politically infeasible. Once water rights became fully tradable; however, there was always the prospect of an increase in demand that would produce a windfall capital gain for those holders of water rights who are willing to sell. It is likely, therefore, that there will be a division of views between rural water users who wish to continue irrigation, and perhaps purchase additional water, and those who are willing sellers. Members of the first group will lose from higher water prices whereas members of the second group will gain. A similar division has been observed in relation to the conversion of farmland to residential use.

In view of the rapid increase in water use until the imposition of the Cap (discussed later) in 1995, there is little likelihood that, if the price of water is set equal to the marginal cost of diversion, aggregate demand will fall short of total available supply. Hence, the equilibrating process will be one in which the price is determined by the intersection of the demand curve with an upper limit to total extractions imposed by public policy. Currently, this limit is set by the Cap, but in future it will be set under the conditions set out in the National Water Initiative.

Policy Uncertainty

Policies regarding the provision, allocation, and pricing of water for irrigation have varied substantially over time. In general, the policy orientation prior to about 1980 represented a developmentalist approach, in which governments took the lead in constructing and financing irrigation works and set water prices, often below operating costs and almost never at a level that allowed a significant return to capital. This is described as Phases 1 and 2 by Musgrave (Chapter 3). As Randall (1981) notes, such policies are commonly adopted in the expansion phase of water management systems, when there is no effective constraint on total water consumption.

More recently, policies typically associated with a mature phase of water management have been adopted. These policies reflect the pressures that arise when extractions approach, or exceed, the capacity of the system to supply water on a sustainable basis. This pressure is reflected in conflict between competing water users, increased concerns about the adverse environmental effects of water management practices, and calls for greater recognition of the claims of the environment generally.

Another manifestation of the advent of the mature phase of water management is the adoption of techniques such as the extraction of groundwater and capture of surface flows, in response to limits on access to in-stream flows. Such actions produce a cycle in which policy responses designed to restrict access to previously unregulated sources of water prompt a search for less accessible, but unregulated, sources, as well as actions designed to exploit weaknesses in the regulatory framework.

In parallel with this process, policymakers and water users are discovering new information about the behavior of the Murray-Darling river system under conditions of stress, and are adjusting and refining estimates of likely future water availability in the light of this information. This process has no natural endpoint, but if patterns of use are broadly constant over some period, and climate change is slow and reasonably predictable, the accuracy of such estimates should improve steadily over time.

In this context, policy uncertainty is inevitable. An attempt to fix policy in advance would imply a failure to adapt to new information. The result would be a steadily increasing divergence between a fixed policy position and the optimal response to changed circumstances, leading eventually to the breakdown of the policy, commonly in a costly and chaotic fashion, as occurred with the Reserve Price Scheme for wool in the late 1980s (discussed later in this chapter).

Some proponents of policy certainty have sought to resolve the problem by specifying, in advance, responses to every conceivable contingency. Although contingency planning is often desirable, the idea that policy certainty can be achieved in this way is illusory. Inevitably, contingencies will emerge that have not been adequately anticipated or to which the planned response turns out to be suboptimal.

The Impossibility of Eliminating Uncertainty

The uncertainties inherent in agricultural production have produced many proposals for policy interventions aimed at eliminating or mitigating it. Over time, however, it has been recognized that the complete elimination of uncertainty is neither feasible nor desirable.

In the debate on this topic, issues of price uncertainty and climatic uncertainty have been closely intertwined. The gradual abandonment of attempts to eliminate price uncertainty set the scene for changes in thinking about responses to climatic uncertainty and the role of irrigation.

The Failure of Price Stabilization

In relation to price uncertainty, Hancock (1930, 66–67) summed up the attitudes that prevailed in Australia for much of the twentieth century:

> The law which [Australians] understand is the positive law of the State—
> the democratic State which seeks social justice by the path of individual
> rights. The mechanism of international prices, which signals the world's
> need from one country to another and invites the nations to produce more
> of this commodity and less of that, belongs to an entirely different order.
> It knows no rights, but only necessities. The Australians have never felt
> disposed to submit to these necessities. They have insisted that their gov-
> ernments must struggle to soften them or elude them or master them.

Governments responded to the demand for greater certainty in agricultural commodity prices by creating a range of price stabilization and underwriting schemes, of which the most notable was the Reserve Price Scheme for Wool, in which a buffer stock was used to fix a minimum world price for wool. The collapse of the Reserve Price Scheme in 1991, which was followed by years of low prices as the stockpile was gradually sold off, contributed to the abandonment of price stabilization as a policy goal, and increasing reliance on market instruments in agriculture generally.

The abandonment of price stabilization influenced discussions of public policy in relation to drought and other forms of climatic uncertainty. The main effect was to increase emphasis on methods by which farmers and other water users could plan for, and adapt to, climatic variability and uncertainty, and to discourage attempts to eliminate uncertainty.

Drought Policy

Until fairly recently, droughts were viewed as exceptional natural disasters, requiring the application of emergency measures. The traditional approach to drought policy in Australia was centered on the administrative policy of drought declaration of districts, normally at the discretion of state governments. A variety of relief measures, which varied over time and between states, was made available to farmers in drought declared areas. Examples included subsidies for the purchases of fodder, low-interest loans, and cash grants. The implicit policy model was that of an unpredictable natural disaster, like an earthquake. Policy was focused on the provision of assistance to farmers who had suffered, or who were exposed to losses as a result of drought.

This policy was criticized by economists including Freebairn (1983), who argued that it undermined incentives to prepare appropriately for drought and encouraged practices such as overstocking. Studies of the implementation of drought relief in the 1980s reinforced Freebairn's arguments and raised new concerns. Only a minority of eligible producers received any relief. In Queensland, 36 percent of the state had been drought declared once in every three years, and over the period 1984–1985 to 1988–1989, about 40 percent of relief had gone to 5 percent of the claimants (Smith et al. 1993).

The main outcome of the Australian debate was the adoption of the National Drought Policy in 1992. O'Meagher (2003) summarizes the key features of the policy:

> Its stated rationale is that "Drought is one of several sources of uncertainty affecting farm businesses and is part of the farmer's normal operating environment. Its effects can be reduced through risk management practices which take all situations into account, including drought and commodity price downturns." The key policy implication is that farmers will have to assume greater responsibility for managing the risks arising from climatic variability. This will require the integration of financial and business management with production and resource management to ensure that financial and physical resources of farm businesses are used efficiently.

The issues surrounding drought policy have been analyzed by Quiggin and Chambers (2004) using a graphical version of the state-contingent production model developed by Chambers and Quiggin (2000). Quiggin and Chambers (2004) showed that the anticipated availability of drought relief would induce farmers to adopt more risky production strategies, such as high stocking rates, and argued that market-based measures, such as rainfall insurance, had the potential to offset risk without distorting production decisions.

Irrigation and Droughtproofing

For much of the twentieth century, irrigation was presented, in combination with relief policies directed at dry land agriculture, as a method of drought-proofing the agricultural sector.[2] The idea is that a guaranteed supply of irrigation water would eliminate reliance on variable natural rainfall, and thereby eliminate losses arising from drought.

As has already been noted, irrigation allows for a more stable supply of water, and can be used to offset fluctuations in natural flows. Nevertheless, the idea of complete stabilization, with fully secure water allocations, does not make economic sense, because either water would be underused in high flow years or investments in storage would be excessive. On the other hand, if farmers design their production strategies to take advantage of high flow years, they will inevitably incur losses in low flows years.

Many of the ideas put forward in the past were revived by the Farmhand Foundation, a group established in 2002 with backing from major corporations such as News Limited and Telstra. The Foundation canvassed ideas such as the large-scale use of polythene piping to reduce evaporation losses and reconsideration of the Bradfield scheme to divert coastal rivers inland. These ideas were subject to vigorous criticism. In particular, the Wentworth Group, a group of scientists, argued against droughtproofing and in favor of a reduction in water use in irrigation. These criticisms had some effect. The chairman of the Farmhand Foundation recently observed (Farmhand Foundation 2005) "While there is no one way to 'drought-proof' Australia, we continue to be reminded the hard way that we live in the driest inhabited country on the planet and that we cannot continue to use water as we have been doing these past two centuries."

Guarantees of Resource Access

In the presence of variable and uncertain supplies of water, or other resources, it is natural for users to seek security of access. It is, of course, possible to guarantee access for some users; however, resource security is, in large measure, a zero-sum commodity. The more security is given to one group of users, the less there is for everybody else.

As is discussed later, well-defined systems of property rights can help to resolve competing demands for resource security by allowing for a variety of contingent trades in rights; however, accurate definition of property rights is only feasible when the availability of the resource is well understood.

Risk Reduction, Risk Allocation, and Risk Management

Risk and uncertainty cannot be eliminated. But there are a variety of measures for reducing uncertainty, and for managing and allocating risk.

Reducing Uncertainty

Uncertainty can be reduced with improved information. In particular, improved methods of weather forecasting can reduce climatic uncertainty. Although there have been significant improvements in short-term forecasting (up to seven days) in recent years, these are of relatively minor importance in the context of irrigated agriculture. Of more interest are seasonal projections of fluctuations associated with the Southern Oscillation (also known as El Niño) and longer-term projections of climate change.

Although the idea of policy certainty is a chimera, uncertainty about policy can be reduced through the adoption of transparent policy processes, based on consistent principles. With such processes in place, policy changes arising from short-term political pressures or from the arrival of new ministers and other office-holders should be minimized. Policy change should arise primarily as the result of the arrival of unanticipated new information.

To achieve such stability, it is important that the establishment of the policy process should be based on clearly argued principles and should achieve a high degree of consensus. Otherwise, changes in the balance of political power are likely to lead to the reversal of decisions imposed by temporary majorities.

Managing Risk

The idea of managing risk is central to modern thinking about uncertainty. It is intuitively obvious that irrigators can adopt strategies that increase or reduce their exposure to particular sources of risk. For example, farmers whose entire area is planted to high-value crops that are highly sensitive to reductions in the volume of water applied will experience more risky flows of income than farmers with a mixture of high-value and more robust low-value crops. Farmers may also employ a range of market-based measures for risk management, such as futures markets and crop insurance.

Although this point is reasonably clear at an intuitive level, it has proved difficult to represent adequately in the context of economic theories of production under uncertainty. The standard approach, based on the concept of a stochastic production function, implies that uncertainty increases linearly with expected output, allowing no real capacity for producers to manage risk.

Chambers and Quiggin (2000) show how a state-contingent representation of production can provide a more realistic representation of active risk management strategies. In addition, this representation allows for an integrated treatment of production-based and market-based risk management strategies (Chambers and Quiggin 2004).

The central idea of the state-contingent approach is to model farmers as allocating resources between different states of nature, corresponding to high and low water availability. Allocation of extra resources to preparation for states of low water availability reduces the risk to output change. It is also possible to model more and less flexible production technologies and assess their implications for the elasticity of demand for water.

Principles of Risk Allocation

The central principle of risk allocation is that risk should be allocated to the party best able to manage it; however, the appealing simplicity of this principle masks some complex and intractable issues. First, some risks are not easily separated and the party best able to manage one risk may not be able to manage other associated risks so well. Second, as noted by Chambers and Quiggin (2004), risk management has both technological and financial aspects. In many cases, farmers are well placed to manage risk in a technological sense, but have limited access to financial markets, and therefore cannot realize the full value of the risk management techniques at their disposal.

The Murray-Darling Basin, the Cap Process, and the Implications for Risk and Uncertainty

Problems of risk and uncertainty have played a central role in the development of policies for the management of the Murray-Darling Basin. Quiggin (2001) presented a summary of developments in policy for management of the Murray-Darling Basin from Federation to the late 1990s, and a comprehensive coverage of historical elements is provided in the preceding chapters.

Environmental problems and competition for water use became evident during the 1970s and acute during the 1980s, signaling the arrival of the mature phase of water management in which the marginal social cost of water use is high and increasing over time. The Cap on new extractions and the accompanying creation of markets for trade in water rights have been major components of the policy response to problems of excess demand for water in the Basin.

The need for water allocations to be transferred between users naturally raised the issue of trade. The argument for trade in water rights is simple and appealing. The market would ensure that the aggregate allocation of water could be capped, and ultimately reduced, without imposing high costs on existing water users. Those who placed a high value on water could buy rights from those whose valuation was lower. The central idea of creating a market for trade in water rights is that rights would be reallocated from low-value uses such as pasture to high-value uses such as fruit and vegetables. Although this reallocation would not, in itself, do anything for the environment, it would reduce the cost and the social and economic dislocation associated with reductions in the aggregate allocation of water.

It rapidly became apparent that this appealing idea was an oversimplification of the water management problem. Water is not a homogeneous commodity. Water in one place, and at one time is not a good substitute for water in another place or at another time. Because it is heavy and bulky, moving water from one place to another or storing it over time is difficult and expensive.

Water is a complex commodity. The structure of rights created by a century of water management is even more so. At the time of the 1994 meeting of the Council of Australian Governments (CoAG), few or no water users possessed property rights comparable to titles to land. The closest approximation was a license to take water, typically attached to a particular piece of land. On the other hand, a great many existing and potential users had expectations that water would be available to them.

As a result, the first problem with water trading was to determine who had water rights. A major difficulty was the emergence of *sleepers* and *dozers*. These were landholders who had water licenses attached to their land, but had never used them (sleepers) or had ceased to use them (dozers). As soon as water became a tradable commodity, the licenses held by such sleepers became a tradable commodity. Because extractions from the Murray-Darling Basin were already at or near 100 percent of natural flows in 1994, it was not possible to allow both the allocation of water to sleepers and dozers and the continuation of existing allocations to users who did not possess guaranteed rights. With some exceptions, the outcome was that users who had been receiving water under various provisions, but who had no specific entitlement, did not receive tradable rights, although sleepers' rights were upheld.

The interaction between poorly specified property rights and the unforeseen significance of sleepers meant that the policy failed to produce the desired outcomes. In particular, although users who were allocated tradable water rights benefited from increased certainty, this was more than offset by the greater uncertainty faced by the remaining users, whose collective claims considerably exceeded the available volume of water.

The National Water Initiative and the Assignment of Risk

By the end of the twentieth century, it was clear that hopes for a rapid transition to a system of fully tradable water rights were misplaced. Many risks and uncertainties remained unresolved. In this context, a new set of proposals was put forward at the 2003 CoAG meeting, which produced an announcement (but not a detailed specification) of a set of policy proposals referred to as the National Water Initiative (CoAG 2003).

Two major principles were announced. The first was that, in future, water allocations should be stated as shares of available water, rather than as specific volumes. This approach deals with fluctuations in water availability by sharing the total amount available among users in proportion to their share. It raises the question of whether it will continue to be possible, as at present, to distinguish between high-security and low-security rights. The difficulties with this approach are discussed by Freebairn and Quiggin (2006).

The second principle, and one particularly pertinent in this context, concerned an approach to the sharing of risk arising from changes in the aggregate availability of water. Under this principle, the risk of changes in water availability due to new knowledge about the hydrological capacity of the system will be borne by users. The risk of reductions in water availability, arising from changes in public policy, such as changes in environmental policy, will be borne by the public, and water users will receive compensation for such reductions.

The principles of the National Water Initiative were elaborated in more detail in a statement issued by the 2004 CoAG meeting (CoAG 2004) and some of this is summarized in the preceding chapter. Importantly, in the context of risk, the Communiqué specified a framework that assigns the risk of future reductions in water availability as follows:

- reductions arising from natural events such as climate change, drought or bushfire to be borne by water users;
- reductions arising from bona fide improvements in knowledge about water systems' capacity to sustain particular extraction levels to be borne by water users up to 2014. After 2014, water users to bear this risk for the first 3 percent reduction in water allocation, the relevant state or territory government and the Australian government would share (one-third and two-third shares respectively) the risk of reductions of between 3 percent and 6 percent; state/territory and the Australian government would share equally the risk of reductions above 6 percent;
- reductions arising from changes in government policy not previously provided for would be borne by governments; and
- where there is voluntary agreement between relevant state or territory governments and key stakeholders, a different risk assignment model to the previously mentioned may be implemented.

The general principles set out in the National Water Initiative are consistent with the approach to risk allocation set out previously. There are, however, numerous problems to overcome.

In the short term, the consensus required to implement a policy of this kind was upset when the Commonwealth government announced, in the lead up to the 2004 election, that its contribution to the National Water Initiative would be funded by the withdrawal of payments to the states previously made as part of National Competition Policy. The Commonwealth's action was a serious breach of the notions of transparent and predictable policy essential in areas of this kind, and increases the likelihood of more opportunistic policy changes in the future.

A second class of problems relates to implementation. For instance, suppose that long-term average rainfall declines in line with the predictions of climate change models. It will be difficult to determine whether the reduction is in fact due to climate change, or merely represents a run of dry years. Although the risk in both cases is supposed to be borne by water users, it is likely that the appropriate response and the resulting allocation of costs between users will differ. Implementation of the NWI principles might therefore be difficult, or even impossible, due to the information requirements of distinguishing between the two types of impacts. Similarly, trade between major areas, which draw from different dams with different reliability/management rules, could undermine the reliability of existing rights. It is unclear whether this would be regarded as a policy-induced change.

A more pressing problem relates to the transition from the current set of water rights, which involves serious over-allocation in many catchments, and over-allocation for the system as a whole, to a more sustainable level of use. Quiggin

(2006) suggests that the adjustment path could be eased, and risk reduced, if governments offered to repurchase renewal rights or, equivalently, make a cash payment to rights holders in return for conversion of their existing rights into fixed term rights, with no option of renewal. In effect, this involves a transfer of risk from holders of water rights to governments.

Conclusion

As new knowledge about water systems emerges and new demands, such as increased concerns about environmental flows, arise, policies must adjust. The adjustment process inevitably creates uncertainty for both new and existing water users. It follows that the allocation of risk and uncertainty is a crucial problem in the design of institutions for water management in Australia.

Risk cannot be eliminated, but it can be managed. Improvements in risk allocation have the potential to yield substantial improvements in welfare. The National Water Initiative has made a promising start, but much more remains to be done.

Notes

1. This approach is being challenged. Details are provided by Edwards in Chapter 10.

2. Additional technological solutions, most notably cloud-seeding, were also investigated without success.

References

Chambers, R. G., and J. Quiggin. 2000. *Uncertainty, production, choice, and agency: The state-contingent approach.* New York: Cambridge University Press.

Chambers, R. G., and J. Quiggin. 2004. Technological and financial approaches to risk management in agriculture: An integrated approach. *Australian Journal of Agricultural and Resource Economics* 48(2): 199–223.

CoAG (Council of Australian Governments). 2003. Communiqué, August 29, 2003, http://www.coag.gov.au/meetings/290803/index.html (accessed August 17, 2004).

CoAG (Council of Australian Governments). 2004. Communiqué, June 25, 2004, http://www.coag.gov.au/meetings/250604/index.htm (accessed August 17, 2004).

Farmhand Foundation. http://www.farmhand.org.au/ (accessed April 15, 2005).

Freebairn, J. 1983. Drought assistance policy. *Australian Journal of Agricultural Economics.* 27(3): 185–199.

Freebairn, J., and J. Quiggin. 2006. Water rights for variable supplies. *Australian Journal of Agricultural and Resource Economics* 50(3): 295–312.

Hancock, W. K. 1930. *Australia.* Brisbane: Jacaranda.

IPCC (Intergovernmental Panel on Climate Change). 2001. *Climate Change 2001: Working Group I: The Scientific Basis.* Geneva: IPCC.

Jones R., P. Whetton, K. Walsh, and C. Page. 2001. Future Impact of Climate Variability, Climate Change and Land Use Change on Water Resources in the Murray-Darling Basin: Overview and Draft Program of Research. CSIRO Atmospheric Research, September.

Knight, F. 1921. *Risk, uncertainty, and profit.* New York: Houghton Mifflin.

O'Meagher, B. 2003. Economic aspects of drought and drought policy. In *Beyond drought: People, policy, and perspectives,* edited by L. C. Botteril and M. Fisher, 109–130. Collingwood, Victoria: CSIRO Publishing.

Quiggin, J. 2001. Environmental economics and the Murray-Darling River system. *Australian Journal of Agricultural and Resource Economics* 45(1): 67–94.

Quiggin, J. 2006. Repurchase of renewal rights: A policy option for the National Water Initiative. *Australian Journal of Agricultural and Resource Economics* 50(3): 425–35.

Quiggin, J., and R. G. Chambers. 2004. Drought policy: A graphical analysis. *Australian Journal of Agricultural and Resource Economics* 48(2): 225–251.

Randall, A. 1981. Property entitlements and pricing policies for a maturing water economy. *Australian Journal of Agricultural Economics* 25(3): 195–220.

Smith, D., M. Hutchinson, and R. McArthur. 1993. Climatic and agricultural drought: Payments and policy. *Drought Network News* 5(3): 11–12.

Thomson, C. 1994. The impact of river regulation on the natural flows of the Murray-Darling Basin. Technical Report 92/5.3. Canberra: Murray-Darling Basin Commission.

CHAPTER 6

The Institutional Setting

Lin Crase and Brian Dollery

T HE HISTORICAL AND LEGAL CONTEXT provided in earlier chapters offers useful and differing lenses through which to consider Australian water policy formulation. An alternative, but compatible, approach is to invoke the conceptualization of institutions, which has emerged as an important strand of research within the economics literature. In this case we borrow North's (1990) definition of institutions as the formal and informal rules that constrain an individual's behavior and consequently determine human interaction. In the interests of completeness, we also endeavor to consider organizations or governance structures as part of the institutional arrangements that pertain to Australian water policy.[1]

Academic interest in institutions partly stems from deficiencies within the neoclassical tradition in economics. More specifically, the neoclassical approach has itself been hampered by the assumed existence of perfect information and a level of individual rationality usually not supported by empirical tests (Wolozin 2002). By way of contrast, analysis of institutions shifts the focus to a comparison between real-world alternatives, and recognizes the bounded rationality of human decisionmaking.

In addition to catering for imperfect information, institutional analysis provides considerable scope for integrating the findings from other disciplines.

Enthusiasm for undertaking institutional analysis of water resource policy in Australia, and elsewhere for that matter, can be traced to several factors. First, institutional analysis provides an elegant means of dealing with the vexing policy debate over the relative merits of markets as a vehicle for allocating resources and state apportionment of water. In the institutional milieu, the question is reshaped around the notion of transaction costs, which ". . . embrace(s) all those costs that are connected with (a) the creation or change of an institution or organization, and (b) the use of the institution or organization" (Furubotn and

Richter 1992, 8). Thus, the question is no longer one of whether markets or governments do a better job of allocating water between competing users—the matter of import is the combination of rules that minimize the costs of decisionmaking.

A second and related factor promoting interest in water institutions is the fugitive and transient nature of the resource and the consequences of this for decisionmaking. Unlike many other goods, water has both public and private good attributes. In addition to the complex bifurcation between the private and public good dimensions of water, the decision over the allocation and use of the resource might be made by entities other than the state or the individual user. More specifically, common property regimes frequently emerge in water resource management where a group is given decisionmaking powers over the management and allocation of the resource. This implies that state, private, and common property regimes all come into play in the management of water. Moreover, as a general rule, these property right regimes are not mutually exclusive: That is, state, common, and private property exist simultaneously in a system of nested rules (Ostrom 1990) or as an institutional hierarchy (Challen 2000, *24–25*). The advantage afforded by the institutional lens for dealing with such complexities has seen expanded interest in its application to water policy analysis.

Third, interest in water institutions can be traced to the progressive realization that engineering solutions to the problems of increasing scarcity are deficient. For example, in the context of irrigation technology, Ostrom (1992) argued that many of the most technically advanced irrigation systems have still been found wanting, primarily because of the inability to design and implement satisfactory institutions.

Fourth, because institutional analysis extends to a consideration of both formal and informal rules, it provides a useful framework for understanding the evolution of water policy in a changing social context. One of the criteria by which to adjudge the effectiveness of institutions is the alignment of formal rules with informal conventions or social norms (Pagan 2003). For instance, formalized markets have been shown to be more effective in cases where competition is the social norm, but tend to be less effective as formal institutions where alternative values circumscribe the decisionmaking process (see, for example, North 1990). In Australia, heightened awareness of, and greater regard for, the natural environment is an important emerging social norm. In addition, the recent past has been characterized by expanded acceptance of the role of the individual and the paramountcy of markets when facing the pressure of global competition (Crase and Dollery 2006). Institutional analysis is well equipped to trace the repercussions of these significant but disparate influences over water policy formulation.

Notwithstanding its advantages as an analytical tool, "[v]ery little attention, particularly in practical policy analysis, has been given to . . . comparative assessment of alternative institutional structures for resource allocation" (Challen 2000, 5). It is the ambitious task of this chapter to partially meet this challenge by providing an overview of important institutional elements that circumscribe Australian water policy.

The chapter itself is organized into three main sections. In section 1 the various elements of Australian water property rights and governance structures are described, highlighting the contrast between important water sectors. Transaction costs and their role in defining superior institutions are considered in section 2. Although the transaction cost approach is a useful conceptual tool, these costs are themselves the manifestation of imperfect information and this matter is also considered in section 2. Some brief concluding remarks are then offered in the final section of the chapter.

Property Rights and Governance Structures of Australian Water

Defining institutions in a manner that encompasses the formal and informal rules of the game that govern human behavior, and the governance structures that give effect to those rules, implies that there is some demarcation between these two elements. In practice, any delineation is often illusory and will vary with the level of analysis and the topic under scrutiny (Saleth and Dinar 2004, 25). The notion of property rights and their relationship to governance structures provides a useful demonstration of the overlap between these two concepts.

Commons (1968, 6) describes institutions as the rules that indicate to individuals what they "must or must not do (compulsion or duty), what they may do without interferences from other individuals (permission or liberty), what they can do with the aid of collective power (capacity or right), and what they cannot expect the collective power to do on their behalf (incapacity or exposure)." This definition provides an important insight into the pivotal role of property rights and explains why property rights are themselves regarded as synonymous with institutions in much of the economic literature (Challen 2000, 18). However, there is also a propensity to use the term *property rights* loosely and this gives rise to confusion on two fronts. First, the purported tragedy of the commons, which Hardin (1968) assumed would attend open access, is often erroneously equated with all divergences from individual or private property (Quiggin 1986). The second main area of confusion relates to the difference between property rights and ownership. A property right provides some authority or power over the management of a resource or object that may or may not accord with the notion of ownership in a legal sense. Put simply, property rights are concerned with the rights of individuals to impose restrictions on the behavior of others.

In order to illustrate this point, Challen (2000, 64) considered the institutional hierarchy that applies to the regulation of surface water use in the Lower Murray and Riverland regions of South Australia and the Riverina and Murrumbidgee regions of New South Wales. In light of the prominence of this region and its influence over Australian water policy, a more detailed analysis of property rights in this setting follows.

Property Rights Within Australian Irrigation

At the highest level of the institutional hierarchy in surface irrigation is a combination of state and federal government agreements that establish the rules for joint regulation of the resource. As noted earlier in this book, the principal agreement in the geographical setting chosen by Challen (2000) is the Murray-Darling Basin Agreement, and it is this Agreement that provides the framework for management of the shared resource as common property. The second tier in the hierarchy is represented by State property and acknowledges the preservation of the Crown's authority over water resources within State-based legislation. Thereafter, rights are assigned to subordinate authorities in either of two forms. First, the state can assign common property rights to entities, like management boards responsible for government irrigation schemes or private irrigation bodies. Secondly, the state can directly allot individual rights to licensed users or riparian right holders. Government and private irrigation schemes can also assign private rights to participant irrigators.

It is important to note that institutional hierarchies of this nature are not restricted to the governance of surface water or the locale of the Murray-Darling Basin. For example, groundwater, which provides for about two-thirds of Western Australia's water needs (WARC 2000, *1*), can also be conceptualized in this manner. Considered as an institutional hierarchy, the upper tier of the governance arrangements in this instance comprises the national and state governments' shared commitment to the National Strategy for Ecological Sustainable Development (1992). In essence, this strategy establishes the guiding principles by which the state body should exercise decisions over the resource. More specifically, the former Water and Rivers Commission[2] developed water allocation plans on the basis that there was provision of environmental water that met the objectives of the National Strategy for Ecological Sustainable Development (WARC 2000). The subordinate tiers in the institutional hierarchy in this case comprise water corporations and other water supply bodies that operate well fields. At this level, common property rights are exercised. At the lowest tier reside individual water users, some of whom gain decisionmaking authority from the common property institution (e.g., urban water customers without direct access to groundwater) whereas others gain authority directly from the state entity (e.g., mining companies extracting the resource directly, private irrigators or households with a garden bore licensed by the state). Regardless of the setting, institutional hierarchies are the norm for the allocation of water resources in Australia.

Considerable attention has been given to the level of attenuation applied to water property rights as part of the transformation of Australian water policy. This has been dominated by debate about the individual rights of irrigators and the impact of attenuation on the productive applications of water (see, for instance, Anderson 2005). In a broad sense, a common theme amongst the various states has emerged in response to calls for improved specification of irrigators' water rights. By and large, the reform effort has focused on reducing the level of attenuation at the lower tiers of the institutional hierarchy and simultaneously clarifying the limits to authority at higher tiers. These changes have taken two main forms: the devolution of the governance of irrigation districts from the state to local entities, and enhanced private property rights for individual irrigators. In addition to these general themes, specific reforms have targeted the quality of title that applies to water resources. Each of these topics is briefly explored in the context of changing social norms before offering some contrasts with the reforms undertaken in the urban water sector.

Strengthening Common Property Rights Within Irrigation Districts. An important element of the CoAG Agreement on Water Resource Policy (Water Reform Framework) of February 1994, and later the Competition Principles Agreement in April 1995 was that water agencies were required to provide efficient water services that separated regulation, service delivery, and water resource management functions. In addition, these arrangements required greater financial self-sufficiency on the part of irrigation (Rigden 1998, 6). In Section 6(g) of the Water Reform Framework the principle that irrigation customers be given greater control over the management of irrigation areas was entrenched and states responded to this in two main ways. First, some states opted to continue state-ownership of irrigation infrastructure while divesting control and responsibility for the management of those facilities to irrigators. This approach is evident in Victoria and Tasmania, for example. The second approach saw states divest ownership of infrastructure to irrigators in the form of private irrigation companies. Instances of this approach can be found in New South Wales.[3] In this case, the state continues to provide only limited financial support in recognition of the degraded condition of infrastructure at the time of transition and as a vehicle for encouraging the adoption of environmentally preferred irrigation technologies. The upshot of both approaches has been similar; a marked enhancement in the exercise of control of water resources as a group resource at the common property level in the institutional hierarchy and a diminution in the role of the state, on a day-to-day basis at least.

Reduced Attenuation of Private Property Rights for Irrigators. In addition to enhancing common property rights within irrigation, the states have generally reduced the attenuation applied at the individual resource level; private property rights have been enhanced. In all states, the nexus between water and land has been broken insomuch as water rights are now generally separable from land and tradable in a market setting; however, state-specific and, in some cases, region-specific constraints remain that constitute attenuation of individual rights, albeit modest by comparison to those applied during the development phases of irrigation. For example, local rules can prohibit the export of all water access rights from some districts and caveats can be applied on the use of water in particular contexts. As McLeod and Warne point out in Chapter 7, exit fees have also been imposed in some irrigation districts. The result of these arrangements is the imposition of some attenuation of individual rights by the group of irrigators, not the state. Often attenuation has been based on the premise that individual action gives rise to undesirable third party effects; however, as recently noted by the Productivity Commission (2006), care needs to be taken to distinguish between those externalities that are pecuniary and those which have stronger grounds.

Notwithstanding the persistence of these forms of attenuation, water policy reform has *prima facie* aimed to enhance the property rights of individual irrigators, albeit generally over a smaller quantum of water. In general, this has occurred on the premise that trade in water will improve water-use efficiency, encourage private investment in water-saving technologies, and simultaneously bring an environmental dividend. In this context, New South Wales has enacted the Water Management Act 2000 while in Victoria the Water Act 1989, and its most recent amendments embodied in the Water (Irrigation Farm Dams) Act 2002, are to be updated by reforms outlined in the White Paper entitled Securing Our Water Future. In South Australia, the Water Resources Act was enacted in 1997. In the Australian Capital Territory, the Water Resources Act 1998 specifies the sequential process for managing water resources and the property rights for water users in Queensland are enshrined in the Water Act 2000. The Water Management Act took effect in Tasmania at the beginning of 2000 and the Northern Territory has amended its Water Act 1992 to comply with the thrust of the reforms. In Western Australia, the state has relied primarily on a series of amendments to the Rights in Water and Irrigation Act 1914 to meet the reform objectives although the Irrigation Review completed in 2005 foreshadows more extensive changes. A review of the legislative frameworks in each state yields two important conclusions. First, the progress against the agreed targets established by the early CoAG process, and later the National Water Initiative, varies considerably among jurisdictions. Second, there are marked differences in the institutional treatment of the various water sectors—a point addressed later in this chapter and given closer attention in Chapters 8 through 11.

It should be stressed that whereas legislative accommodations have often been made on the premise of enhanced irrigation efficiency, evidence of the wider impacts of these changes is mixed. More specifically, universal claims of enhanced efficiency need to be treated cautiously. For instance, as part of the process of enumerating the benefits of stronger individual water rights and the accompanying enthusiasm for water markets, a trial interstate water trading project was established in the irrigated areas in New South Wales, Victoria, and South Australia between Nyah and the mouth of the River Murray. This pilot project commenced in 1998 and by June 2001 had facilitated the permanent trade of almost 14 gigalitres of water, mostly to South Australia. Notwithstanding that the official review of this project claimed that all water had unequivocally been moved to a higher value use (Young et al. 2000), the reality was that almost all of this water was sleeper or dozer rights (Crase et al. 2003). The spillover diminution of the rights of other irrigators was not considered in the official review of the project and, as Quiggin has already noted in the preceding chapter, this might be better conceptualized as a zero sum game (at best). Accordingly, some grounds remain for contesting the long-term efficacy of this approach (Crase et al. 2003).

Enhancements in the Quality of Title. Clearly, there are several elements of the water resource allocation to irrigators that have yet to be fully clarified—return flows and water quality represent only two of these—but *prima facie* individuals in this sector enjoy greater control of the resource than they did in the preceding phases of water development. Greater decisionmaking power that now resides with the lower levels of the institutional hierarchy in irrigation has also been accompanied by progress to enhance the information that accompanies those decisions.[4]

All states have progressively moved toward a system of volumetric entitlements for water use in irrigation. This is no trivial accomplishment in light of the formerly disparate and incompatible arrangements across states. Similarly, legislative and administrative changes are underway to convert groundwater rights into a volumetric portion of sustainable yield, although the cases examined by McKay in Chapter 4 provide sobering evidence of the difficulty of this task in over-allocated systems. Nevertheless, progress on this front is important as it provides individual decisionmakers with a more robust vehicle for adjudging extraction, and thereby enhances the operation of water markets. In addition, it provides regulators with an improved vehicle for monitoring and enforcing compliance.

However, the conversion of rights to volumetric allocations still embodies significant variations between, and in some instances, within states. For example, in New South Wales, the state with the largest irrigation sector, a system of prioritized rights has been adopted. Thus, some irrigators hold stronger claims than others. Rights have also been unbundled into access and use components in this state. A share entitlement provides the holder with a portion of accessible water after the sustainable limits have been defined, whereas an extraction entitlement allows entitlement holders to withdraw water at a designated location, subject to specific conditions. In many respects these arrangements are a manifestation of attempts to more clearly define dimensions of water use which were largely ignored in the rush to create water markets in the early CoAG reforms. More specifically, use rights can be used to limit some spillover effects, like salinization.

In Victoria, irrigators now hold water shares allocated in a manner that reflects water available for consumption (i.e., similar to the previous water rights in this state), and 80 percent of what was previously termed *sales water* (i.e., formerly regarded as surplus but not tradable) is to be specified as a secure but low-reliability right and will now be tradable. Thus, irrigators in this state hold two different types of right. Victoria has also foreshadowed the development of a system of shares in delivery capacity (e.g., channel capacity) and a license is required for use on a particular site. Delivery shares offer scope for enhancing the efficient timing of irrigation water, particularly in congested systems. Perhaps ironically, they may also provide a vehicle for reassigning at least one component of irrigation water to the land. For instance, Goulburn-Murray Water, the largest irrigation entity in Victoria, has foreshadowed the introduction of delivery shares. These shares attach the right (and cost) of accessing adjoining infrastructure to the land. As a result, an irrigator who sells all of their access rights will still be required to share in the cost of infrastructure, so long as they retain ownership of land and it continues to abut irrigation infrastructure. Exit fees, payable by those who no longer wish to continue in the irrigation industry, have yet to be determined. However, set at the proposed levels (between $214 and $490 per megalitre, depending on the irrigation district) they are likely to represent a significant consideration for potential sellers in a market where access rights are presently sold for around $1,200–$1,500 per megalitre. In effect, these arrangements represent an attempt at the community level to wind back the unbundling of land and water rights, a process commenced as part of the early CoAG reforms.

Notwithstanding policy nuances between states, individual irrigators now have greater clarity concerning their individual rights. Some communal attenuation of individual rights endures, although this varies significantly between jurisdictions and is also attracting the attention of authorities vested with the responsibility of promoting national competition (e.g., the Australian Competition and Consumer Commission).

An important ingredient of the broader water reforms has been the acknowledgement of the rights of the environment, or in an anthropocentric sense, the rights of those individuals who value environmental amenity. Earlier in this book the imposition of the Cap on extraction within the Murray-Darling Basin was described (see McKay in Chapter 4). This occurred in response to information available to the upper tiers of the institutional hierarchy that pointed to the deleterious consequences of the continued growth of water use; however, recognition of the requirements to retain water in-stream to satisfy environmental demands has not been limited to the Murray-Darling Basin. All states have now enshrined environmental rights in legislation and established priority claims in order to secure ecological benefits.[5] At the catchment level this has manifested itself in an array of water management plans (Tasmania and New South Wales), stream-flow management plans (Victoria), local water management plans (South Australia), water allocation and management plans (Western Australia and Queensland), and water resources management plans and environmental flow guidelines (Australian Capital Territory). In essence, these arrangements have been put in place to give effect to the prior claims of the environment over consumptive activities. They have also become an important mechanism by which irrigators' rights are defined insomuch as withdrawals for irrigation must lie within the scope of those plans.

Notwithstanding the reservations expressed by McKay (Chapter 4) about the legal standing of irrigators' rights over a variable resource, and Quiggin's (Chapter 5) observation that there is some ambiguity about the assignment of risks between the state and individual irrigators, there remains some disquiet about the apparent trade-offs struck by policymakers to gain concession for environmental purposes. Here we contend that irrigators now enjoy stronger individual rights which are less attenuated by the state, albeit over a smaller consumptive pool. What was previously state property has been vested in individuals and common property irrigation entities. By many measures, there has been a substantive redistribution in favor of the irrigation sector because the value of those rights has been patently enhanced by the establishment of water markets and, in time, this benefit can be realized through trade.

This part of the reform process has itself been plagued by mixed messages that confuse environmental outcomes with gains in economic efficiency. Attempts to thoroughly define individual irrigators' rights, and thereby constrain the rights of others, has led to increased pressure on those elements of the resource that are presently less clearly articulated. First, the right to return flows from irrigation activities has not been clearly specified in any jurisdiction. Consequently, the implementation of technological efficiencies at the farm level, which have frequently been espoused as bringing a water-saving or environmental benefit, has eroded return flows. Farmers who save water can simply redeploy that water to other productive and extractive uses. In this case the purported savings are nonsensical. Previously, these flows bestowed increased reliability on downstream users and, in some cases, underpinned environmental flows. The Productivity Commission (2006, 39) cites work by Engineering Australia which showed that increasing irrigation efficiency from 80 to 90 percent in the Riverland district in South Australia resulted in a 22 percent reduction in groundwater inflows into the adjacent River Murray. One method of encapsulating this component of water use is to specify water entitlements as net volumes—that is, any reduction in return flows is considered to be a component of the water entitlement. Although this approach has been employed elsewhere (e.g., western U.S.), to date the costs of assembling the data necessary to specify Australian water entitlements in net terms has been regarded as unjustified.

Second, because the rights to water quality remain largely ill-defined, there is scope for some users to overlook this dimension of the resource. In this context it should come as no surprise that the aforementioned official review of the interstate water trading pilot project was ultimately forced to concede that "in the long run, interstate trading can be expected to have a negative impact on river salinity" (Young et al. 2000, 3). Some constraints on trade have subsequently been applied along the River Murray to account for the likely high-impact and low-impact effects of a given trade. Differential levies are applied to trade in an effort to capture these third party effects, although the data needed to establish these arrangements universally is limited. An alternative approach requires the use of carefully specified use rights, but the same data gaps and the costs of accumulating information limit the usefulness of this approach.

From an individuals irrigators' perspective, the structure of water rights is *prima facie* more amenable to market transactions and, theoretically at least, should yield improvements in the allocation of the resource between productive uses. However, as noted in the context of the interstate trading project and more generally by Quiggin in Chapter 5, the increased value of rights has been accompanied by the activation of sleeper and dozer entitlements in some states. This behavior stands to undermine the environmental claims on the resource and potentially offset some of the effort to provide greater quality of title for irrigators.

Of particular interest in this context is the evolution of these formal institutions and their relationship to the informal norms within Australian society. On one hand, Australians would now appear to place greater store in the value of environmental amenity. Moreover, the availability of more information about the degraded state of the environment has accompanied this transformation. This trend is not unique to Australia, however, and it is common in many countries that enjoy sufficient income to value future benefit over expanded current production. The reformation of Australian water policy, such that changing social norms pertaining to the environment are now embodied into formal institutions, provides evidence of the influence of informal institutions over the policy formulation process.

By way of contrast, however, there is also evidence to suggest that there has been a metamorphism of Australian society on another front. Australians now, perhaps more than ever, appear to have greater confidence in the capacity of the market, and the individual within the market, to deliver efficacious outcomes in a variety of contexts. The power of organized labor has diminished, contracted labor is now commonplace, market arrangements for service delivery by government agencies is the norm, and the need to sustain international competition is seldom challenged. It can be argued that the devolution of decisionmaking to the lower tiers of the water institutional hierarchy, and the attendant enhancements in the rights of irrigators, is a manifestation of the evolution of formal water institutions in line with this wider social inclination. What remains to be seen is whether both of these trends can remain congruous with the formal institutions being developed to meet these challenges.

Property Rights for Urban Water Users

Institutional reform of the urban water sector in Australia has not been immune to the two broad social themes described earlier. In common with the reforms in irrigation, the consequences of social change can be traced to adjustments in the property rights enjoyed at the different levels within an institutional hierarchy. In order to illustrate this approach, a comparative hierarchy developed for urban water in Victoria by Crase and Dollery (2006) is reproduced as Figure 6-1.

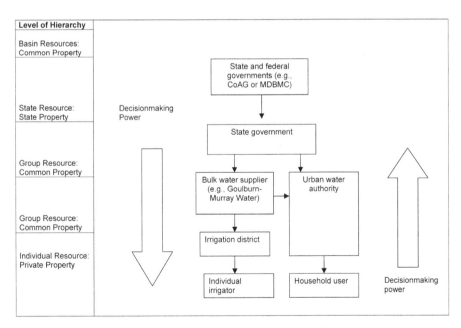

Figure 6-1. *Institutional Hierarchy for a Set of Urban and Irrigation Water Users in Victoria*

Source: Adapted from Crase and Dollery 2006.

Although the reform of water policy has altered the property rights within the urban water institutional hierarchy, the policy approach adopted in this instance varies markedly from that employed in irrigation. Unlike the reformation of water policy in irrigation, only modest emphasis has been placed on the property rights that circumscribe the lowest tier of the urban institutional water hierarchy; the urban water user. Moreover, in some instances, there has been a deliberate effort on the part of policymakers to ensure that the rights of urban water users are significantly attenuated relative to their irrigator cousins. Crase and Dollery (2006) adapted Scott's (1989) earlier property rights typology to examine the relative attenuation of rights at the individual and group level for both irrigation and urban sectors in Victoria. In general, they found that the property rights at the individual irrigator level are far less attenuated than those of urban water users. In addition, there is some evidence that the rights of individual irrigators may be considered superior to those of urban water authorities that hold common property rights on behalf of urban communities (Crase et al. 2005). Partially at least, this can be explained by the intense efforts on the part of the irrigation sector to secure enhanced individual property rights in exchange for the improved status of environmental claims.

Institutional reform in the urban sector has been punctuated by an expansion of the states' attenuation over individual's rights to water. For instance, the National Water Initiative offers four core elements to encourage greater urban water reform. First, the standard requirement for "continued implementation of full-cost recovery pricing for water in both urban and rural sectors" (CoAG 2004, 1) is repeated from earlier CoAG initiatives. However, the remaining ingredients focus on "actions to better manage the demand for water in urban areas." Second, there is to be "a review of temporary water restrictions," presumably with the aim of amending their status to a more permanent footing. Third, "minimum water efficiency standards and mandatory labelling of household appliances" are to be applied with the aim of enhancing consumer awareness of water use. Finally, "national guidelines for water sensitive urban design" are to be developed (CoAG 2004, 1). Once again, this approach implies an attenuation of urban users' rights by mandating that households use water in a manner strictly acceptable to the state.

Although the rights enjoyed by those at the lowest tier of the urban hierarchy have been subject to greater attenuation, this does not apply unilaterally to those entities that hold common property on behalf of urban water users. In some states, this level of the hierarchy is in the form of local government while in others, regional or metropolitan entities occupy this tier. State-based reforms have, in some instances, treated these entities in a similar manner to the common property enjoyed by management boards of government irrigation schemes or private irrigation corporations. For example, in Victoria, urban water authorities hold bulk entitlements that, to all intents and purposes, are equivalent in status to those held by irrigation schemes (albeit much smaller in volumetric terms). By way of contrast, in New South Wales, urban water rights vested in local governments have superior status to all irrigation claims, including those irrigators who hold high security access shares.

However, to assume that urban authorities enjoy similar or superior rights to irrigators would ignore the important political influences exerted over the exercise of those rights. At the national level, the parliamentary secretary to the prime minister's office has declared that "cities must learn to use the water they had (sic) more efficiently before they considered buying irrigation water from outside their catchments" (Nairn 2005, cited in Quiggin 2005, 1). At the state level, the urban reforms are characterized by mandated water restrictions, compulsory water-saving targets, and punitive measures against recalcitrant customers. Moreover, whereas urban authorities have access to the water market, unfettered participation is either administratively or politically curtailed (see, for instance, Crase et al. 2005). By way of contrast, in irrigation, administrative and political impediments to market participation are largely on the decline.

The divergent approach to property rights in the urban and irrigation sectors raises some interesting questions about the policymakers' perceptions of the distribution of the informal norms described earlier: that is broad enthusiasm for the market ethos and stronger demands for environmental enhancements. The fondness for stronger property rights at the lower level of the institutional hierarchy in the irrigation sector, accompanied by simultaneous tighter regulation of urban water users, might suggest that Australian water policymakers believe that these two changing social norms are asymmetrically distributed between urban and rural Australians. Put simply, the current policy response suggests that urban Australians place greater emphasis on environmental outcomes and are prepared to allow their rights to be attenuated to achieve these goals.[6] In contrast, rural Australians with an interest in irrigation are more easily swayed by the case for greater economic efficiency in irrigation. Unfortunately for policymakers, the bifurcation between rural and urban Australians is more problematic and the likelihood that these institutional arrangements will endure as water becomes increasing scarce in urban contexts is doubtful.

Transaction Costs and Information in Australian Water Institutions

"One of the key premises in the institutional economics literature is that institutional change occurs only when the transaction costs of reform are less than the corresponding opportunity costs of doing nothing" (Saleth and Dinar 2004, 13). The earlier sections in this chapter traced considerable reform of Australian water policy and placed this in the context of two distinct social phenomena: a thrust toward greater emphasis on environmental amenity, and widened acceptance of the role of atomistic behavior in markets to deliver efficiency. Moreover, we contend that the evident reforms are themselves a manifestation of the inability of the former institutional arrangements to harmonize with these two norms.

Like Saleth and Dinar (2004), Challen (2000) considers the process of institutional reform in the context of transaction costs. However, he diverges from Saleth and Dinar's description of the process by identifying two broad genres of transaction costs. The first is termed *static transaction costs* and comprises the range of costs associated with the current institutional arrangements. Prior to the reform of irrigation, this might encompass the costs of state officials gathering information and deciding on a water allocation or infrastructure project on behalf of irrigators in government-owned irrigation districts. Static transaction costs would also include any efficiency losses that emanate from the assignment of an inefficient allocation of the resource on the part of the irrigation bureaucracy. The second genre of costs is termed *dynamic transaction costs* and this category consists of transition costs and inter-temporal opportunity costs. The former of these dynamic costs arise from the legacy of institutional history. For instance, in irrigation, the decision by state governments in the development phase to allow irrigators expansive use of the resource as if it were individual property has severely constrained any serious consideration of alternatives other than less attenuated individual rights in this sector. A corollary of taking action to amend institutional arrangements is that future choices may then be unavailable, or at least more costly to achieve. This represents the second category of dynamic transaction costs; namely, inter-temporal transaction costs.

The point of unbundling transaction costs in this manner is to recognize that preferred institutional arrangements are defined by the set of rules that minimize overall transaction costs (i.e., the sum of static and dynamic transaction costs). Moreover, because social norms change over time, the benefits and costs accruing to an institutional form will also change. For instance, as social norms alter, static transaction costs might rise because the existing institution is unable to acceptably approximate the marginal social benefits and costs of an activity. Similarly, when additional information becomes available, the true cost of the institutional status quo might be revealed, thereby prompting reform.

However, altering these arrangements to reduce static transaction costs will itself be costly, particularly if property rights have to be divested from the few with intense preferences to the many with less intense preferences (Horn 1995). In contrast, institutional history suggests that devolution of property rights from the dispersed many to the concentrated few has relatively low costs, because the intense preferences of the few encourage them to mobilize political resources to secure such a re-distribution.

In the context of these broad observations about institutional dynamics, it is possible to shed some light on Australian water reform. First, the devolution of property rights in the institutional hierarchy for the irrigation sector can be traced to several events. The institutions that formerly allocated resources in this sector had been found wanting on efficiency grounds. State owned and managed irrigation trusts were characterized by inefficient allocation of the resource and a culture that emphasized a use-it-or-lose-it approach to water distribution. These phenomena were clearly out of step with the wider reformation, occurring within the economy that emphasized smaller government, efficient markets, and greater influence by users and customers over decisionmaking. These events were accompanied by the historically strong claims of irrigators over water use, regardless of the legal specification of the resource as state property. Thus, the static transaction costs of inefficient government allocation have combined with the dynamic costs of choosing institutions that might further attenuate individual rights. The impact of these events has been the assignment of stronger rights for irrigators under formal institutions.

Second, the transfer of decisionmaking power upward within the institutional hierarchy for the urban sector can be accounted for within this framework. Urban users have previously enjoyed only modest attenuation over their use of the resource. For example, metering water use in urban areas is a relatively recent practice in many cities. Notwithstanding the liberal nature of these rights, water use has hitherto played only a small part in urban life. Data published by the ABS (1999) reveals that household expenditure on water (comprising water and sewage rates and charges) represents about 0.8 percent of total weekly expenditure by the average household. Moreover, this is less than one-third the amount expended on alcoholic beverages and one-twentieth of the expenditure on transport. In this context the transfer of rights upward in the institutional hierarchy in this sector engenders only modest dynamic transaction costs. In addition, the re-allocation of greater decisionmaking to the state may accord with lower static transaction costs if the state is able to access and employ superior information relative to individual urban users. More specifically, state water authorities may be better equipped than individuals to adjudge those measures necessary to secure the supply of urban water. In this instance, the welfare loss associated with mandatory restrictions and the like may be less than the welfare loss that accompanies unbridled individualism in urban water consumption. This is particularly the case if urban water authorities are unwilling or unable to use price to communicate information to consumers. In this context, the mobilization of property rights to higher tiers within the urban institutional hierarchy accords with a net reduction in transaction costs.

Conclusion

The framework employed to analyze institutions in this chapter points to the inevitability of future reform in Australian water policy. Changing social norms continue to challenge the acceptability of the institutional status quo, and particularly the tolerability of attenuated property rights at different levels within institutional hierarchies.

Australia's water institutions are typified by a hierarchical distribution of property rights covering a variety of state, common, and private property regimes. The development era that preceded recent reforms saw private rights severely attenuated in the most profligate water sector, irrigation. This approach was a function of the wider goals associated with irrigation expansion, in the form of policies promoting regional development, soldier settlement and the like; however, the reformation of irrigation institutions has been characterized by less attenuation of individual irrigator's rights in all states. This is expected to bring significant efficiency gains in this sector in the form of enhanced trade and investment to achieve more sparing use of the resource. It has also been heralded as one of the major institutional accomplishments of water reform in this country (see, for instance, *The Economist* 2003; Saleth and Dinar 2004, *177*).

Nevertheless, pronouncing the success of this approach appears premature. In some instances the efforts to enhance irrigators' rights over a smaller consumptive resource pool have led to pressure on other elements of the water cycle that are less clearly defined—the challenge of preserving return flows by adequately defined rights being a case in point.

In contrast to the irrigation sector, acknowledgment of water scarcity has resulted in institutional adjustments in the urban sector that have clearly increased the attenuation of individual rights. We have contended that this is a function of the transaction costs of pursuing alternative arrangements in this setting at this point in time.

However, the Australian economy and society of the twenty-first century differs markedly from that which saw the paramountcy of agriculture in the development generation, and that which has provided irrigators with such strong claims on the resource. Significant differences have emerged on both social and economic fronts such that the wider status of agriculture is itself under challenge. The divergent institutional arrangements between irrigation and urban sectors seem likely to attract intense scrutiny as water becomes increasingly scarce in both sectors. The durability of the current arrangements is thus questionable and the response of policymakers may prove instructive, particularly for observers of institutional dynamics.

Notes

1. There is insufficient scope in this chapter to cover all definitional aspects of institutions. For an excellent treatment of the various dimensions of institutions see, for instance, Saleth and Dinar (2004, 23–45).

2. In 2004, a new Ministry of Water Resources was formed which consolidated all water policy issues into a single portfolio. Accordingly, the Water and Rivers Commission was subsumed into this portfolio.

3. The privatization process is described in some detail by McLeod and Warne in Chapter 7.

4. This matter illustrates the observation that "there is a two-way relationship between knowledge and institutions, and, in this sense, knowledge and institutions may be viewed as substitutes" (Saleth and Dinar 2004, 24).

5. Prior to these reforms, environmental claims were commonly residual to all other demands. In this context, fresh water flowing to the sea was often regarded as wasteful under the development ethos.

6. Edwards provides a more comprehensive justification for the attenuation of urban users' rights in Chapter 10.

References

ABS (Australian Bureau of Statistics) 1999. *Household expenditure survey 1998–99.* http://www.abs.gov.au/ausstats/abs@.nsf/94713ad445ff1425ca25682000192af2/62bc7fe0325f40e3ca256f7200832ef0!OpenDocument (accessed November 30, 2005).

Anderson, J. 2005. Australian Water Summit: Address to the Australian Water Summit. http://www.ministers.dotars.gov.au/ja/speeches/2005/AS4_2005.htm (accessed May 11, 2005).

Challen, R. 2000. *Institutions, transaction costs, and environmental policy institutional reform for water resources.* Cheltenham: Edward Elgar.

CoAG (Council of Australian Governments). 2004. *Intergovernmental Agreement on a National Water Initiative.* Canberra: Council of Australian Governments.

Commons, J. R. 1968. *Legal foundations of capitalism.* Madison, WI: University of Wisconsin Press.

Crase, L., and B. Dollery. 2006. Water rights: A comparison of the impacts of urban and irrigation reforms in Australia. *Australian Journal of Agricultural and Resource Economics* 50(3): 451–462.

Crase, L., B. Dollery, and J. Byrnes. 2005. An Inter-Sectoral Comparison of Australian Water Reforms. Paper presented at Markets for Water: Prospects for WA. Perth, Australian Agricultural and Resource Economics Society. Perth, September 2005.

Crase, L., B. Dollery, and M. Lockwood. 2003. Water down property rights for the sake of the environment: A consideration of the environmental benefits of attenuated water rights in New South Wales. *Australasian Journal of Environmental Management* 10(1): 25–34.

The Economist. 2003. Survey: Liquid assets. 368:13.

Furubotn, E., and R. Richter 1992. The new institutional economics: An assessment. In *New institutional economics,* edited by E. Furubot, and R. Richter. London: Edward Elgar, 1–32.

Hardin, G. 1968. The tragedy of the commons. *Science* 162: 1243–1248.

Horn, M. 1995. *The political economy of public administration: Institutional choice in the public sector.* Cambridge, MA: Cambridge University Press.

North, D. 1990. *Institutions, institutional change and economic performance.* Cambridge, MA: Cambridge University Press.

Ostrom, E. 1990. *Governing the commons: The evolution of institutions for collective action.* New York: Cambridge University Press.

Ostrom, E. 1992. *Crafting institutions for self-governing irrigation systems.* San Francisco: ICS Press.

Pagan, P. 2003. Measuring Institutional Performance: Understanding Issues in Indian Irrigation. Paper Presented at Australian Centre for International Agricultural Research (ACIAR) Conference, Beechworth, Victoria, July 2003.

Productivity Commission. 2006. *Rural water use and the environment: The role of market mechanisms.* Melbourne: Productivity Commission.

Quiggin, J. 1986. Common property, private property, and regulation: The case of dryland salinity. *Australian Journal of Agricultural Economics* 30(2/3): 103–117.

Quiggin, J. 2005. Water dilemmas flow on. *Connections—Farm, Food, and Resources Issues* July: 1–2.

Rigden, T. 1998. Regulating for a Sustainable Irrigation Industry in Western Australia. Paper presented at the Australian National Committee on Irrigation and Drainage Annual Conference, Sale, Victoria, August 1998.

Saleth, R. M., and A. Dinar. 2004. *The institutional economics of water: A cross-country analysis of institutions and performance.* Cheltenham: Edward Elgar and the World Bank.

Scott, A. 1989. Conceptual origins of rights based fishing. In *Rights based fishing,* eds. P. Neher, R. Arnason, and N. Mollet. Dordrecht: Kulwer Academic.

WARC (Water and Rivers Commission). 2000. *Water facts.* Perth: Water and Rivers Commission.

Wolozin, H. 2002. The individual in economic analysis: Toward psychology of economic behaviour. *Journal of Socio-economics* 31: 45–57.

Young, M., D. H. MacDonald, R. Stinger, and H. Bjornlund. 2000. *Inter-state water trading: A two year review.* Canberra: CSIRO Land and Water.

CHAPTER 7

Coping with the Reforms to Irrigated Agriculture

The Case of Murray Irrigation

Jenny McLeod and George Warne

MOST OF THE WATER EXTRACTED FROM dams, rivers, and aquifers across Australia is used for irrigated agriculture—as much as 67 percent, according to the Australian Bureau of Statistics (ABS 2004). This varies considerably between states, from 40 percent of the available water in Western Australia up to 79 percent in South Australia.

As the nation's largest water user, irrigated agriculture has always borne the brunt of water reforms from state to state. In eastern Australia that reform has come not only from state governments, but also through the intergovernmental bodies such as the Murray-Darling Basin Ministerial Council (MDBMC), and more recently through the Council of Australian Governments (CoAG) as a result of changing Commonwealth water policy and the altered status of the Commonwealth in water affairs.

Australia's largest river system is the Murray-Darling. It crosses four states and has dominated water policy development in Australia. The scope of the Basin in geographic and agricultural terms was described in an earlier chapter, but it is worth noting that as much as 70 percent of Australia's irrigated agriculture is within the Murray-Darling Basin. Irrigated agriculture in Australia takes up less than 1 percent of the agricultural area but produces 28 percent of total agricultural production (CSIRO 2005).

As identified earlier in this volume by Musgrave (Chapter 3), changes to water policy have directly impacted irrigation on two broad fronts. First, governments have been reluctant to continue to financially underwrite the operation of irrigation systems. This has resulted in an increased focus on cost-recovery, higher water prices, and amended institutional arrangements to devolve greater responsibility to irrigators. Second, increased national prosperity has changed the community priorities from those which drove the development of irrigation (drought proofing, closer settlement, increased national agricultural production, and food security) to ones of improved environmental amenity. As the most expansive water user, the irrigation sector has keenly felt these claims for resource re-allocation to satisfy environmental demands.

In addition to specific water related reforms, the irrigation sector faces many of the challenges associated with agriculture generally. Variable returns for outputs, declining terms of trade, and the demise of the family farm are just as prevalent in irrigation as they are in dry land agriculture. Compounding these problems is the social malaise in rural communities that invariably accompanies significant economic adjustment.

Water policy and reform involving the River Murray have been particularly complex and dominated by politics and competing parochial interests. This chapter considers the impacts of changing national priorities for water on irrigators through the experiences of Australia's largest irrigation company, Murray Irrigation Limited, which is based on the River Murray in New South Wales (NSW). It also considers how the community of irrigators represented by the company has responded to those changes. Notwithstanding that the irrigation sector is more extensive in NSW than the other Australian states, this chapter provides detailed insights into the ramifications of water reform. Many similar experiences are evident in irrigation outside NSW.

Geographical Perspective

Murray Irrigation's area of operations was originally developed by the NSW government as the NSW Murray Irrigation Area and Districts. The region is in the southern part of the Murray-Darling Basin, bordered by the River Murray to the south and the Billabong Creek (a tributary of the Murrumbidgee River) and the Edward River (an anabranch of the River Murray) to the north and west (Figure 7-1). It covers an area of 748,000 hectares, almost all commandable using only gravity to deliver irrigation water. Approximately half this area is irrigated in any one year.

The landscape is generally flat with a fall of 1:20,000. Agricultural production in the region includes rice, winter cereals (such as wheat, barley, oats, and canola), dairying, and other livestock production for beef, lamb, and wool. There is also a small, but extremely productive, annual horticulture industry producing processing tomatoes and potatoes. With the exception of drought years the region produces 50 percent of Australia's rice crop, 20 percent of the milk sourced from NSW, 75 percent of NSW's processing tomatoes and 40 percent of the state's potatoes. The farm gate production based on Australian Bureau of Statistics local government area data is approximately $300 million (CSIRO 2005). Rice represents the major commodity, producing 40 percent of the gross value of production, followed by winter cropping (22 percent), dairy production (15 percent), nondairy livestock (14 percent), and fruit and vegetables accounting for only 6 percent. Table 7-1 gives indicative production data for the region during a normal allocation year. Importantly, the agricultural mix is broadly reflective of the dependence on general security water entitlements in

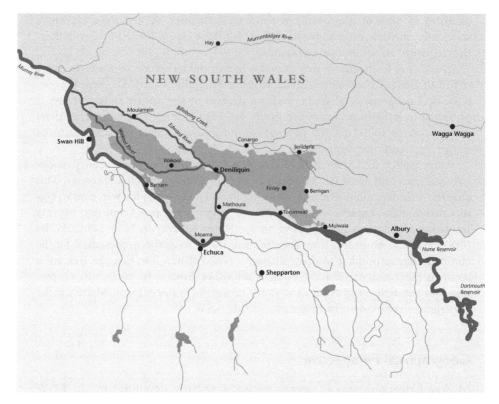

Figure 7-1. *Murray Irrigation Limited's Area of Operations within the Murray–Darling Basin*

Source: Murray Irrigation Ltd. 2005b.

the area of operation. General security entitlements have a relatively low level of surety, varying markedly with the water available in storage. Accordingly, this water allocation regime is more conducive to annual crops or enterprises where other inputs can be substituted during years when there are low allocations.

Significant amounts of local produce are processed within the region. Deniliquin is home to the largest continuously operating rice milling plant in the southern hemisphere. Processing operations greatly enhance the value of agricultural industries to the community, primarily by providing employment opportunities.

The region has a population of approximately 25,000 people. Deniliquin is the largest town with a population of 8,000, and the urban communities in the region are ostensibly reliant on agricultural industries. Irrigated agriculture is a vital part of the local economy and community fabric. It provides employment and drives demand for service industries. In contrast to dry land farming communities across Australia, those populations based on irrigated agriculture have demonstrated greater resilience and capacity to maintain a community fabric, despite the pressures of farm input costs and declining terms of trade for the farm sector.

Table 7-1. *Nominal Production Data for MIL Area of Operations During a Normal Allocation Year*

Summary	Hectares	Tonnes	$Value
Summer Cropping—Rice	82,894	764,866	162,118,616
Summer Cropping—Other	10,706	23,358	4,881,162
Fruit/Veggies	2,591	91,182	24,408,821
Winter Cropping	192,295	1,310,023	90,212,788
Pastures & Hay	620,281	82,065	9,039,220
Dairy	N/A	N/A	62,111,747
Livestock	N/A	N/A	56,835,772
Total	908,767	2,271,494	409,608,126

Historical Perspective

The historical background to the development of irrigation in this region has been detailed earlier by Musgrave (Chapter 3). However, disputes over the River Murray effectively delayed the development of irrigation in the NSW Murray region for 30 years, compared to development in other parts of NSW, South Australia, and Victoria. It was not until 1933 that the NSW government finally began building what was to become the state's largest irrigation network, known as the Murray Irrigation Area and Districts.

Construction continued until 1964, with gravity-fed earthen channels diverting the waters of the River Murray to droughtproof the region—providing reliable stock and domestic water to a vast region regularly devastated by drought. (It is one of four similar systems built by the NSW and Victorian governments in the southern Murray-Darling Basin from the early 1900s to the 1950s, diverting water from rivers including the Murrumbidgee, Murray, Goulburn, and Campaspe). State-based restrictions in NSW ostensibly prohibited the development of irrigated horticulture or permanent plantings of fruit crops or viticulture from being undertaken in the irrigation district.

Earlier bans on rice growing in the Murray Valley were gradually lifted across the Murray area and districts between 1942 and 1967, and for many years irrigators were encouraged to expand their water use. Drought in the mid 1960s provided the first warning of the Murray's resource limitations. For the first time the Murray system was unable to supply the increasing demand along its length and NSW moved quickly to introduce a volumetric licensing system to more effectively account for and control water use in its irrigation districts. Volumetric allocations were introduced for private irrigation schemes in NSW in 1975 and for river pumpers in 1981 (Martin forthcoming).

Privatization

In the early 1980s NSW irrigators were faced with significant government water price increases to cover the operating costs of the distribution scheme. Irrigators in the Murray districts believed that an insufficient portion of the revenue collected was being returned to the region; the maintenance of local infrastructure was being neglected; and there was also a perception that there were significant opportunities to improve both the financial and operational efficiency of the system. Irrigators began agitating for local management, separate financial accounting, and independence from the remote government head office in Sydney.

Irrigation leaders' aspirations for local management were driven by recognition of the obvious opportunities for efficiency enhancements on the financial, water delivery, and service improvement fronts. The benefits to government of privatizing the irrigation network included reduced future liability for: (a) the ageing irrigation supply infrastructure, (b) environmental risk associated with irrigated agriculture, and (c) economic viability of the irrigation supply and drainage system. In addition, separation of commercial water supply functions allowed government to focus on regulation and compliance, often described as government avoiding being both gamekeeper and poacher.

The support of government leaders, given the many complex legal issues involved, was also essential. The proposal for privatization of government-owned irrigation in NSW coincided with a major government policy shift which resulted in the CoAG 1994 policy statement, details of which were presented in McKay (Chapter 4).

The privatization proposal was not as simple as a sell-off of a government asset, but rather a transfer of ownership from government to the irrigators supplied by the system. Establishing sound institutional arrangements for the operation of the new company was fundamental to meeting both the government's and the new company's requirements for risk management, accountability, and control.

Individual leaders representing irrigators and government were instrumental to the privatization process, working closely to develop institutional arrangements and agreements which met the interests of both irrigators and government. Nevertheless, there were a number of potential deal stoppers that included ownership rights of Murray Irrigation Limited's main irrigation supply channel, the Mulwala Canal, the model for ownership of the Water Access License, and government funding for asset renewal of the neglected irrigation supply infrastructure.

Murray Irrigation was ultimately created as an unlisted public company, which fundamentally altered the legal status of the organization. The new entity was accountable to company, trade practices, taxation,[1] and employment laws.

The development of a financial structure capable of meeting the new company's core businesses responsibility of water delivery was essential, including long term modeling taking into account best- and worst-case resource scenarios. Armed with an understanding of the future infrastructure liability of the new entity, irrigators were able to negotiate significant government funding over the first fifteen years of the new company to upgrade neglected infrastructure.

As part of the privatization process, the government annexed the Murray Land and Water Management Plans (LWMPs) to the company's operating licenses. These plans were being developed in conjunction with irrigators in the region at the same time as the privatization negotiations. The plans were being developed in response to rising water tables which threatened the productivity of the landscape. They are a thirty-year natural resource management strategy which links to the company's licenses and thus creates a vested interest in ensuring the success of the plans. The annexing provided an institutional structure with incentives for compliance with environmental initiatives, and has leveraged considerable landholder investment to improve management practices. Arguably, the approach may have proven less successful if not for the extensive involvement of irrigators at the outset.

Privatization required the establishment of new institutional structures to facilitate the development and maintenance of relationships between Murray Irrigation and the NSW government, and between Murray Irrigation and its customers. Presently, Murray Irrigation is an unlisted public company, and the irrigators supplied by the system are exclusively its shareholders. The company is a monopoly provider, and individual irrigators have no option other than to remain shareholders. Share ownership brings with it responsibilities and certain rights—particularly to a defined share of the allocation or annual yield from Murray Irrigation Limited's water access license.

Murray Irrigation's Relationship with Government

Licensing

Murray Irrigation is closely linked to the NSW state government through its licensing arrangements, including licensing for water supply. The company was issued with a bulk water license for approximately 1,450,000 megaliters[2] of general security water from the NSW Murray regulated river system. The company also has a fifteen-year Irrigation Corporation Water Management Works license with the NSW government for the diversion and delivery of water to Murray Irrigation shareholders. In addition, the company holds an Environment Protection License with the NSW Department of Environment and Conservation. These licensing arrangements impose a coordinated discipline on the company and in turn its shareholders. In effect, the licensing arrangements endeavor to force shareholders, individually and collectively, to take a responsible approach to environmental management in the Murray Irrigation region and to minimize any impacts on the downstream environment.

Water Supply

Access to water is determined by the NSW Department of Natural Resources, under the Water Management Act 2000 (NSW) according to the Water Sharing

Plan for the NSW Murray and Lower Darling Regulated Water Sources. As the Murray is a shared water resource, NSW access to water is regulated by the 1992 Murray-Darling Basin Agreement.

Water Pricing

Bulk water is purchased from state water, and bulk water charges are regulated by the NSW Independent Pricing and Regulatory Tribunal (IPART). Bulk water prices are between 20 and 22 percent of the company's operating costs (Murray Irrigation Limited 2005a) and have risen significantly faster than the consumer price index since the company was formed in 1995. This reflects the wider water policy changes advocating full cost recovery.

Funding Deeds

Murray Irrigation is also linked to government through funding deeds for deferred maintenance and Land and Water Management Plan (LWMP) performance. Government funding for deferred maintenance and LWMPs is audited annually by an independent auditor appointed by the NSW government. Annual income from government for the LWMPs is in the order of $7–8 million, with the Commonwealth and state governments contributing equally to the implementation of the Murray LWMPs.

Relationship Between the Company and Its Shareholders

On formation, irrigators serviced by the company were issued shares in the company and entitlements to Murray Irrigation water. The shares represented ownership of the company's infrastructure and assets. The water entitlements represented a share of Murray Irrigation's bulk water license.

The leading irrigator driving privatization established the company's Memorandum and Articles of Association; this document contained the rules for the operation of the company. It also contained the company's water transfer rules and water supply contract which specified the conditions for water supply to the shareholder customers.

Embedded in the Memorandum and Articles of Association were concepts designed to preserve the position of irrigators as it had existed with the state government prior to privatization. The company Memorandum and Articles of Association include the following important principles:

1. Murray Irrigation provides water at least cost and is required to establish and maintain reserves to ensure the ongoing viability of the business.
2. Shareholders have to own land in the Murray Irrigation area.
3. Shareholders may not permanently transfer from their landholding more than 40 percent of the water entitlements applicable to that landholding at privatization.

4. A permanent transfer from the company license will not be approved where the consequence of the transfer would be to have transferred more water out of the corporation than would have been transferred in since privatization.
5. Member directors are elected from geographic regions equating to the former irrigation districts.
6. The right to vote at company meetings is restricted to one vote per landholding.

The creation of Murray Irrigation and the issuing of Murray Irrigation shares and water entitlements created a new institutional structure, with the separation of the land and water assets. Nevertheless, limitations within the company's Memorandum and Articles of Association placed significant caveats over the extent to which water entitlements could be permanently traded from land. Arguably, the rules originally employed by Murray Irrigation reflect the evolution of the firm from its earlier traditions as a government bureaucracy with a social as well as economic function. It is these rules which are now under intense scrutiny as a result of recent water policy reform—a matter given greater attention later in this chapter.

Achievements from Privatization

Staff efficiency—The introduction of technology, improved work flow, and analysis of work performance, particularly in the day labor and maintenance area, has allowed Murray Irrigation to dramatically reduce work numbers from approximately 300 prior to privatization to 130.

Water supply efficiency—Achieving water savings by reducing water losses from the irrigation system is vital. Murray Irrigation has introduced radio control of supply channels as an important step toward improving water supply efficiency. In 2000 Murray Irrigation returned 30,000 megaliters to the NSW government for environmental flows for the River Murray in exchange for capital funding from government for infrastructure investment. This form of quasi exchange is considered later by Crase and O'Keefe in Chapter 11. In addition, since 2000, Murray Irrigation has increased water supply efficiency by about 2 percent for a given NSW state allocation announcement. This is over and above the transfer of 30,000 megalitres to the NSW government.

Nevertheless, improving water supply efficiency has not been without consequences. To improve water supply efficiency the company has tightened its operational criteria for water supply, with subsequent constraints now applied to the flexibility of the water supply service provided to irrigators.

Full cost recovery—Murray Irrigation's annual water charges are set to ensure the business collects sufficient funds to meet all annual operating and refurbishment costs and to collect funds for future asset replacement. Substantial funds are also set aside to build a reserve for years when water sales are low. Full cost recovery has been achieved with only small increases in the bulk water price, once the impact of state bulk water price increases and inflation are accounted for. The collection of funding for future asset replacement is unprecedented in the Australian water sector and meets one of the objectives of the water reform agenda. Murray Irrigation also receives annual funding for specific deferred maintenance works as part of its separation agreements with government. This funding is in addition to revenues received from water sales, but its expenditure is quite specific and approved by the government's program auditor.

Geographic information systems (GIS)—Murray Irrigation has built a complex GIS that covers the area of operation. The GIS provides the company and shareholders with a wide range of information which can be used to accurately locate company assets and property boundaries, help plan infrastructure management, measure rice areas, and map watertables.

One of the most significant changes to environmental management since separation from government has been the development of the company's environment program which includes a mixture of incentives and penalties.

The principal driver for the development of the Murray LWMPs was the significant threat of broad-scale landscape degradation resulting from rising watertables. In a bottom-up approach involving extensive landholder consultation, working groups of farmer and community representatives developed the plan strategies, with support from government and agency staff, in particular NSW Agriculture and the Department of Water Resources. The LWMPs aim to achieve a balance of environmental, social, and economic outcomes acceptable to landholders, the broader community, and government. They are an integrated strategy of farm and district-scale works, education, monitoring, and research and development programs with a thirty-year, $498 million, community-government funding program (Marshall and Norwood 2003). Table 7-2 clearly shows the non-trivial financial commitment by landholders to the plans and the emphasis on on-farm expenditures against the goals of the plans.

Since plan implementation began, the area of land threatened by high watertables has declined from 84,000ha in August 1997 to 3,800ha in August 2005. Murray Irrigation has argued that this has been due to the successful

Table 7-2. *Funding Split for Land and Water Management Plan—Thirty-Year Budget*

Spending Category	Landholder Contribution ($M)	Government Contribution ($M)
Channel seepage	1.7	1.6
Administration	2.7	0.9
Research and Development	1.8	1.8
Education	3.3	3.3
Monitoring	5.6	1.8
Sub-surface drainage	20.0	4.0
Surface drainage	74.0	52.0
On-farm works	258.0	38.0

implementation of the LWMPs accompanied by drier climatic conditions (Murray Irrigation Limited 2005b).

Since commencement of the plans, landholders have contributed $351 million to LWMP activities and the government has contributed $67.54 million. In 2004–2005 a total of $8.5 million of government funding and $5 million from landholders was spent on implementing LWMP activities where incentives were available. Based on large, statistically valid, and independently audited surveys, landholders spent another $38.9 million on LWMP activities that do not attract subsidies (Murray Irrigation Limited 2005a).

In addition to these laudable achievements on the environmental front, a range of other initiatives apply at the shareholder level which have garnered enhancements. These include:

- Rice water policy to ensure rice is only grown on suitable, impermeable soils.
- Total farm water balance policy which limits total farm water use. This policy limits irrigation intensity to four megaliters per hectare or up to six megaliters per hectare where farm best management practices have been implemented.
- Wetland watering on private land. Since 2001/20002 Murray Irrigation has participated in a partnership with the NSW Murray Wetlands Working Group Inc. to deliver environmental water to dry wetlands and natural depressions in the region that have not been flooded for a number of years.

Trade—Although government policymakers appear preoccupied with developing markets for the trading of permanent water entitlements, the irrigation sector has been dominated by annual water trading which has provided irrigators with significant opportunities to improve the profitability of their farm businesses. Murray Irrigation shareholders are Australia's most active net importers of water in the annual water market, often trading in excess of 80,000 megaliters of water annually into the district from connected systems. Significant annual trade between shareholders is also common. By way of contrast, in the ten years since privatization only 3,000 megaliters of entitlements have been traded into or out of the company's operational area on a permanent basis. Permanent trade in water between shareholders is also relatively modest. In 2002–2003, the lowest annual allocation until that point was announced and the importance of the annual water market was clearly demonstrated. About 36 percent of the water available for use by shareholders was sourced from outside Murray Irrigation, including a 124,000-megaliter advance of water from Snowy Hydro Ltd. through a commercial arrangement. At the time of writing (late 2006), farmers in the Murray are faced with the prospect of an even dryer season than 2002–2003. Annual allocations are being progressively downgraded and water trade was temporarily suspended in the Murray and Murrumbidgee Valleys.

Since 1998–1999 Murray Irrigation has operated a water exchange between August and May for shareholders seeking to purchase or sell annual water. Since the concept was first established by a local irrigator organization, the Southern Riverina Irrigation District Council, the exchange has developed into a daily exchange with shareholders able to purchase water, twenty-four hours a day, seven days a week using Murray Irrigation's water ordering telephone line.

The exchange also provides an important Internet-based information service offering data on current prices, volumes on offer, volumes sold, and price trends. Since 2001/2002 buyers from outside Murray Irrigation have also been able to purchase water on the exchange. In 2002–2003, a low allocation year, 60,418 megaliters were sold on the exchange with a market value of over $12 million. Water was traded for prices ranging from $100 to $350 per megaliter, setting a new record price for annual water sales at that time. There is no commission or administrative fees charged on water exchange transactions.

The expanded use of the annual water market, combined with the continuous operation of the water exchange, has highlighted the commercially sensitive nature of information relating to water availability. In this context the state government has been forced to recognize that allocation announcements contain commercially sensitive information. The state has responded by ensuring the confidentiality of the allocation announcement prior to public release, and introduced predetermined dates for allocation announcements.

Current and Future Reform Challenges

The transformation of Murray Irrigation into a commercial, customer-focused business has coincided with a major shift in government and community attitudes toward irrigated agriculture. The privatization deeds signed with government have been enduring. In contrast, changing government agendas focused on limiting water diversions for agriculture and increasing environmental flows for rivers continues to shift views of the irrigation–environment balance. This has highlighted the institutional weakness inherent in the company's water license. In addition to the climatic factors that affect water availability, the company's share of the available water resources has been subject to change without compensation by government. A significant risk to the company is its inability to directly control changes to water sharing rules that influence the availability of a key business input, water.

At the same time that the company has faced four of the lowest ever recorded annual allocations in the region's history of irrigated agriculture, governments have implemented comprehensive water reforms. Government water reforms in the early 1990s initially focused on user pays water pricing and full cost recovery for bulk water services. However, government attention has unequivocally shifted from pro-irrigation development to conservation, with an emphasis on redistributing water from irrigated agriculture to environmental flows. Since 1993, when the MDBMC announced the allocation of 100,000 megaliters each year for the Barmah/Millewa forest for environmental purposes, Murray Irrigation has faced a succession of changes to the way water is allocated between irrigation and the environment.

Ministerial Council Cap on Diversions from the Murray-Darling Basin

As noted in earlier chapters in this book, one of the most pronounced changes in water policy occurred in 1995 when the MDBMC introduced a Cap on diversions from the Murray-Darling Basin at the 1993/1994 level of extraction. In the NSW Murray, the long term Cap represents 87 percent of original entitlements.

When the Cap was introduced, river pumpers and those in horticultural districts in NSW used between 55 percent and 64 percent of their licensed entitlements. In contrast, Murray Irrigation's ten year annual water use was 110 percent of license entitlement. This occurred because MIL farmers would take up unused allocations and the NSW government had historically allowed this to occur (say, in the form of off-allocation flows). The change in government allocation policy was designed to stop growth in water use in the NSW Murray Valley by limiting the maximum announced allocation. Another important change was the introduction of a carryover policy allowing individuals to carryover unused water to the next year and reducing access to supplementary water or off allocation water. All general security license holders were treated the same way, irrespective of their historical water use.[3] As a result, in a normal irrigation year Murray Irrigation irrigators are reliant on water trade to maintain their pre-Cap water use. The government's Cap implementation policy resulted in conflict between government and irrigators with the ensuing debate focusing on whether the Cap should be introduced on the basis of history of use or licensed entitlement. The upshot is that all entitlements were treated equally regardless of the history of use. Arguably, from the perspective of those irrigators who had fully employed their water entitlement, this amounted to treating all water users equally badly.

Environmental Flows for the Snowy River

In 1998 the NSW government held the Snowy Water Inquiry as part of the corporatization process for the Snowy Mountains Hydroelectric Scheme. The Inquiry was run by the NSW government because NSW is the largest shareholder in the Scheme. The Inquiry was concerned with environmental problems arising from the operation of the Snowy Scheme. It was to provide a range of fully costed options to balance the needs of competing users of water, including the environment. The final report from the Snowy Water Inquiry was delivered in October 1998 and recommended returning 15 percent of average natural flow to the Snowy River. In December 2000 agreement was reached between the NSW, Victorian, South Australian, and Commonwealth governments to restore 28 percent (350,000 megaliters) of average annual natural flows to the Snowy River below Jindabyne. The agreement included $375 million to acquire water savings, $75 million of which is for environmental flows in the River Murray. The intent of the government's heads of agreement over the Snowy is to fund efficiency savings and operation improvements so environmental water can be delivered without negative impacts on the irrigation industry. The expectation of irrigators is that water savings projects will fall short of the required water (a view supported by Crase and O'Keefe in Chapter 11) and governments will enter the water market to meet their commitments.

Water Management Act NSW and Water Sharing Plans

Between 1999 and 2002 the NSW government embarked on major changes to water legislation and water administration. In December 2000 the Water

Management Act was passed, replacing the Water Act of 1912. The Water Management Act is a key element of the NSW government's compliance with CoAG's 1995 water reform package. It required governments to establish clearly defined property rights in water for both water users and the environment, and the separation of land from water. The development of clearly defined and certain entitlements to water was a highly contentious aspect of the government's new water legislation, irrigators arguing the Act did not provide a water right with the certainty required for investment.

Between 1999 and 2004 the NSW government set about developing valley Water Sharing Plans across the state in consultation with valley water management committees, appointed by government. The Water Sharing Plans were required as part of the government's new Water Management Act 2000. They apply for ten years and describe in detail how water will be shared between water users and the environment and also between different consumptive users (e.g., general security and high security license holders). The Water Sharing Plan for the NSW Murray includes an enhanced Barmah/Millewa forest allocation, changes the way Menindee Lakes are operated, and addresses successive changes to allocation policy that the NSW government had introduced since the Cap was announced. The expected impact of the Murray Water Sharing Plan is to reduce diversions to 3.8 percent below the Cap.

A key weakness of the Water Sharing Plans and also the Water Management Act from the perspective of irrigators is that of diminishing certainty of water security once the plans are enacted. There is no indication or boundaries to potential changes within the Water Sharing Plans at the expiration of their ten year term.

Living Murray—Environmental Flows for the River Murray

Since the late 1990s the MDBMC has driven momentum for increased environmental flows in the River Murray. This process was re-badged in July 2002 as "The Living Murray" and involved the MDBC analyzing the benefits and impacts of providing an additional 350,000; 750,000; or 1,500,000 megaliters to the River Murray as environmental flows. It has caused considerable uncertainty for irrigators and related communities. A 1,500,000-megaliter redistribution would be equivalent to about a 20 percent reduction in water use in the NSW Murray, assuming proportionate contributions from the Goulburn, Murrumbidgee, and Murray systems. Murray Irrigation has aimed to be actively involved in the Living Murray and has funded a study by RM Consulting Group that quantified the economic impacts of water withdrawal on Murray Irrigation shareholders. The company also funded a review of the three major scientific reports that have underpinned the science behind the Living Murray.

National Water Initiative (NWI)

NSW irrigators' property rights were redefined by an open ended water license and a ten-year Water Sharing Plan (WSP) under the Water Management Act of 2000 and the Water Management Amendment Act of 2004. The WSP provided some certainty that access to water would not be reduced administratively without compensation during the life of the plan; however, it provided no guarantees at the end of the WSP. Irrigators were dissatisfied with the strength of the property right to water provided by legislation, arguing it was not sufficient to underpin ongoing investment in irrigated agriculture.[4]

In addition, the alleged need for increased environmental flows, particularly in the River Murray, was also attracting national interest and fueling uncertainty. Real and potential threats of reductions in the volume of water available for irrigated agriculture combined with low allocations caused by extended drought fueled conflict between governments and irrigators. These threats also galvanized the support of regional business leaders and the banking industry and culminated in political recognition at both state and Commonwealth levels of the national imperative of establishing a transparent water property rights framework supported by legislation.

Widespread conflict between governments and rural and regional communities about the attenuation of irrigators' rights and failure to legislate for clearly defined and certain rights caused the Commonwealth government to be more directly involved in water policy. In 2003 the NWI was announced. Thereafter followed eleven months of extensive debate between Commonwealth and state governments along with a range of diverse groups covering irrigation, environment, and business interests. Eventually, the NWI initiative was signed by the prime minister and premiers of most Australian states. Irrigators hold high expectations of the NWI; they are particularly hopeful that it will provide license holders with more certain water rights; however, it also demands changes to Murray Irrigation's permanent water trading rules. This is described in detail in the following section.[5]

Ironically, the increased uncertainty has coincided with a time of rapidly increasing values in water entitlements (Murray Irrigation Limited 2005c). Possibly, this simply relates to a period of widespread shortage of water through drought and the market judging the future of water supply as one with less water available to farmers.

Expansion of Permanent Water Trading

Disparity between NWI-induced expansion of permanent water trading and Murray Irrigation's trading rules needs to be addressed by the company. Changes to the company's permanent trading rules also require the support of 75 percent of shareholders voting at a general meeting of the company.

The NSW government has recently passed the Water Management Amendment Act of 2004 that will impose significant penalties on Irrigation Corporations that do not allow permanent trade in entitlements out of the Irrigation Corporation in accordance with the NWI. Under the NWI, Murray Irrigation is being asked to change its rules so that anybody can own water entitlements. Buyers will not need to own or use land within the Murray Irrigation operating area or, in fact, to own or use land at all.

Currently, water held outside of irrigation schemes can reside with any individual. A new register of water ownership is also being created in NSW.

As an independently tradable item with a recent history of rapidly rising capital value and potential annual income, water entitlements are expected to become more easily transferable. They will also invariably provide greater security for lenders, and potentially attract investor interest (D. Grant, pers. comm.).

Permanent Trading Options and Stranded Assets

Murray Irrigation share and water entitlements can be sold to other Murray Irrigation shareholders or non–Murray Irrigation shareholders, externally. When water is sold to an external buyer it is deducted from Murray Irrigation's bulk water license.

Murray Irrigation's current rules prevent any sale of water entitlements which would result in the number of water entitlements on its license falling below the sum held at the time of privatization. The NWI requires Murray Irrigation to change this rule to immediately allow up to 4 percent of the company's entitlements to be available for permanent transfer out of the area each year with no constraints from 2009. This equates to an initial maximum of 57,000 megaliters of access rights per year.

Murray Irrigation has commenced the process of identifying what changes are required to allow expanded permanent trade in line with the NWI. Change is also required to avoid Murray Irrigation's water entitlements being devalued relative to entitlements held elsewhere.

By definition, Murray Irrigation's business includes extensive infrastructure with ongoing maintenance and refurbishment costs. One of the implications of significant outward water trade is the threat to the continued viability of its business. The major challenge confronting any irrigation company facing a significant alteration to its share and water entitlements structure is to ensure that, although water might be traded from the area of operations, there is a continued obligation to contribute to the running costs of the business. This is euphemistically referred to as the stranded assets problem, where residual irrigators are forced to carry the costs of irrigation infrastructure across a smaller shareholder base. Murray Irrigation has advocated tagging of permanently traded water to its current source. In essence, a purchaser of MIL entitlement would be buying a water access right attended by the conditions imposed at its source. For instance, if a caveat was placed over the access to MIL entitlements in the MIL region, that same caveat would apply even if the access right was held in another location. The tagging of water entitlements with an annual fixed charge is viewed as a potential mechanism to protect the company's investments. Tagging has recently received government support, with foreshadowed NSW legislation and an intergovernment agreement.

Currently, two mechanisms are being explored to facilitate expanded permanent water trade between states and between valleys: trade based on tagged entitlements or trade based on exchange rates. Irrigation corporations in NSW have declared tagging as their preferred mechanism for expanding permanent trade rather than exchange rates because of the uncertainty regarding water rights, changing government policy, and the third party impacts of exchange rates (McLeod 2004).

Water trade based on tagged entitlements results in the buyer owning entitlements that are the same, and remain the same, as the entitlements being sold. In contrast, water trade based on exchange rates effectively converts any water purchased into another product. For buyers who are also water users, this will be a water entitlement with the characteristics of other water entitlements at the receiving location. Once converted, the water entitlements permanently reflect the seasonal allocations and reliability of the receiving location, despite the fact the water may come from storages with very different reliabilities. This requires the development of exchange rates to convert the seller's water entitlements to the buyer's water entitlements. It is important to note that both tagging and exchange rates represent practical responses to one of the dilemmas created by water trading—namely, the impact of trade on water supply reliability. This is an area requiring considerable policy refinement.

Any consideration of change to the company's Memorandum of Association to allow expansion of permanent trade also raises important questions about the objectives of Murray Irrigation. For example, to what extent can a company of this type retain a not-for-profit philosophy, seeking to supply water at least cost to its shareholders? In an environment of expanded permanent trade, irrigation entities must compete to attract and retain investors. The current structure and institutional apparatus within such entities may no longer be sustainable. Put simply, the direction and governance arrangements which evolved from preceding state bureaucracies may not withstand this focus on competition and trade.

However, the requirement to gain shareholder support for proposed changes to the company's trading regimes cannot be underestimated. Shareholders have a strongly held view that Murray Irrigation and their communities are vulnerable to permanent loss of water if trading rules are relaxed. This is despite the relatively small number of permanent water entitlements sold from within the area of operation since privatization (less than 0.02 percent of entitlements) and the extent of inward annual water trade.

The company has recently recommended a package of changes to its members to allow the expansion of trade in entitlements, consistent with the National Water Initiative. These changes include:

- Allowing individuals who do not own land in the Murray Irrigation area of operation to own Murray Irrigation Water Entitlements;
- Allowing individual landowners to sell all but five Water Entitlements from a landholding; and
- Introducing an exit fee on entitlements when water entitlements are transferred from Murray Irrigation's Water Access License.

The package of changes are intended to ensure Murray Irrigation's rules: (a) comply with the law; (b) protect the viability of the company; and (c) provide increased flexibility in the ownership of land and water assets to shareholders and investors.

Murray Irrigation's costs are largely fixed and independent of the volume of water sold. The company's charging policy historically relied on collecting approximately half its costs from a fixed water entitlement charge and half its costs from water usage charges. Murray Irrigation's rationale for allowing nonlandowners to own Murray Irrigation entitlements is straightforward—it maintains an income stream from the fixed charges and also maintains the characteristics of Murray Irrigation entitlements, irrespective of whether the annual yield on the entitlements is used in the Murray Irrigation area of operation.

The introduction of an exit fee where Water Entitlements are transferred from Murray Irrigation's license is designed to protect remaining customers from increased charges resulting from lower entitlement volumes. Murray Irrigation considered introducing an annual fixed charge, associated with land-based irrigation infrastructure, referred to as access fees. This option was rejected because it was considered likely to significantly increase the risk of bad debts where entitlements were transferred from a landholding, but fixed annual charges remained. This option would have also required that amendments be made to the company's charging policy. One of the outcomes Murray Irrigation is seeking from the constitutional change is to minimize the impacts on those members who choose not to modify their business operations.

Crase and Dollery (Chapter 6) noted earlier that the imposition of exit fees has been attracting the attention of the Australian Competition and Consumer Commission—the entity charged with monitoring anticompetitive behavior. However, advocates favoring the removal of exit fees appear to have ignored the responsibilities of directors to ensure the financial viability of the company and the legislative requirements of NSW irrigation corporations to accumulate sufficient funds to ensure the long-term viability of the irrigation infrastructure. This infrastructure is characterized by a lumpy expenditure profile. In sum, mandating against exit fees stands to seriously undermine the financial viability of communal irrigation infrastructure.

Conclusion

Murray Irrigation, like most irrigation entities in Australia, has undergone a radical transformation in response to water policy reform. The devolution of responsibility to shareholder irrigators has realized significant benefits on several fronts. First, significant technical efficiencies have emerged since privatization. Second, environmental achievements within the area of operation subsequent to the privatization of the infrastructure have been commendable. Third, the privatized entity has successfully harnessed the benefits arising from some of the reforms—the encouragement and expansion of annual trade to ameliorate water shortages being a case in point. Fourth, the company has been able to develop competencies that assist in representing the interests of its shareholders to all levels of government, including having the capacity to research and analyze issues and provide constructive input to the state and Commonwealth governments on an extremely challenging and complex area of public policy—water policy.

The company has recently recognized the need to do much more to develop a value for the company, its staff, its assets, and the role it plays—if the shareholders see themselves as customers but not shareholders there is a risk that this conflict will undermine the future repair, maintenance, modernization, governance, and growth of the company and the region it serves.

However, water reforms have also brought significant challenges. The disposition of governments to seek increased environmental flows will invariably challenge the ingenuity of irrigators and irrigation companies like Murray Irrigation. Particularly pressing is the conflict between encouraging free and open trade and the responsibility to retain irrigation infrastructure at reasonable cost to irrigators. More generally, the pace of reform shows no signs of slowing and the irrigation sector is unlikely to avoid the costs of future policy changes.

Notes

1. Taxation issues have been significant for the company and in contrast with the tax equivalents regime, which the state purportedly applies to government-owned irrigation supply authorities, these provisions are largely being ignored now.

2. Changes to NSW water legislation in 2000 and 2004 and the enactment of the Water Sharing Plan for the NSW Murray Lower Darling Regulated River Water Sources in 2004 altered Murray Irrigation's water licensing arrangements. The company has since been issued with Water Access Licenses under the Water Management Act, including licenses for general security irrigation, high security irrigation, and urban conveyance and supplementary water.

3. Apart from some interim measures to provide preferential access for historically high users to excess stream flows, during the past five years this access provision has proved virtually useless from a production perspective. The difficulties of complying with the Cap have also been highlighted by Quiggin (Chapter 5) and McKay (Chapter 4), although from a state-wide perspective.

4. There are divergent views about the strength of property rights now held by irrigators. Prior to the Water Management Act (2000) all water could be taken from irrigators insomuch as it was solely vested in the Crown and simply licensed to users. However, as noted by McKay in Chapter 4, a history of provision and state policy that encouraged use led irrigators to justifiably believe they had a strong right to water. The political implausibility of taking water confirmed these views. Thus, from an irrigator's perspective the design of WSP represents a diminution in security of entitlement.

5. As briefly noted earlier, the NWI was supplanted in some respects by the Commonwealth governments' 2007 push to have states in the Murray-Darling Basin refer powers.

References

ABS (Australian Bureau of Statistics). 2004. Water Account Australia, Cat no. 4610.0 http://www.abs.gov.au/Ausstats/abs@.nsf/b06660592430724fca2568b5007b8619/9f319 397d7a98db9ca256f4d007095d7!OpenDocument (accessed 1st October 2006).
CSIRO (Australian Commonwealth Scientific and Research Organization). 2005. *Irrigation in perspective: Irrigation in the Murray and Murrumbidgee Basins, A bird's eye view.* Canberra: CSIRO.

McLeod, J. 2004. Water trading—Tagging or exchange rates? *Irrigation Australia* 19(4): 23.

Marshall A., and C. Norwood. 2003. Murray land and water management plans. *Natural Resource Management* special edition (June), 2–8.

Martin, W. Forthcoming. *Water policy history on the Murray River*. Deniliquin, NSW: Southern Riverina Irrigators.

Murray Irrigation Limited. 2005a. *Murray Irrigation Limited Annual Report*. Deniliquin, NSW: Murray Irrigation Limited.

Murray Irrigation Limited. 2005b. *Murray Irrigation Limited Sustainability Report*. Deniliquin, NSW: Murray Irrigation Limited.

Murray Irrigation Limited. 2005c. Permanent Water Trade History. http://www.murrayirrigation.com.au/watexch/permhist.php (accessed 5th October 2006).

CHAPTER 8

Hydroelectricity

Ronlyn Duncan and Aynsley Kellow

GOVERNMENT ATTEMPTS TO REFORM AUSTRALIA'S energy and water sectors in the past decade have occurred during a time of sustained change in the agricultural industry. The confluence of these factors has heightened demand for limited water resources. Corporatization of publicly owned utilities and the restructure of Australia's electricity industry have introduced a radically new management regime for hydrogenerators. Having renounced their capacity-driven mindset, hydrogenerators are now reluctant to surrender any part of their water allocation unless it is economically attractive to do so. In addition, water reforms have driven a fundamental shift in water management with the requirement for environmental flows and opportunities for water trading. This chapter examines the relationship between the energy and water sectors in the context of the restructured Australian electricity industry.

It is important to note that water used to generate hydroelectricity is usually categorized as nonconsumptive because when it has passed through the turbines of a hydroelectric power station, it remains in-stream for other users and its use does not represent a large deficit to a waterway or storage system (ABS 2004). Indeed, the main findings from Australia's Water Account for 2000–2001 were that water consumption in the electricity and gas supply industry, predominantly for hydroelectricity generation, was 1,688 gigaliters (ABS 2004, 3). This figure represents only 7 percent of Australia's water consumption compared to 16,660 gigaliters for agriculture (67 percent of Australia's total consumption), and 24,909 gigaliters for the entire nation. A rather different picture is painted, however, when in-stream use is taken into account. In this case water use for hydroelectricity generation rises to 49,244 gigaliters (ABS 2004, 9). Tasmania's use is highest with 37,405 gigaliters, followed by Victoria at 4,479 gigaliters, and then New South Wales and the Australian Capital Territory combined at 4,118 gigaliters. The latter set of figures reflects

Tasmania's reliance on hydroelectricity and large-scale hydrogeneration in the southeastern states.

Consequently, it is important to consider in-stream figures in conjunction with consumption figures, as a reliance only on the latter would paint a distorted picture of water availability in catchments where hydrogenerators have control over the timing and volume of water flows. Not surprisingly, tensions arise when water stored or used for electricity generation is not available when it is needed by other users. For example, in Tasmania most rainfall occurs in the winter months yet demand for water is highest in the summer months, particularly from farmers wanting to irrigate crops during the short warm growing season (DPIWE 2001, 8). In-stream availability and Tasmania's hydro storages are at their lowest in the summer months. A similar situation exists on mainland Australia—electricity demand is highest in the hot summer months, yet it is the summer months when primary production also requires water for irrigation. Hence, high demand in the energy sector coincides with high demand in the agricultural sector. Within this context, it could be argued that although water is nonconsumptive in hydroelectric power generation, if it is not available when needed by other users, it might as well be consumptive. The previously mentioned in-stream figures highlight the influence hydrogenerators can wield over other users and are a sobering reminder of the importance of the hydroelectricity sector to water policy reforms and outcomes. They are also a testament to how crucial it is for amicable, equitable, and revisable arrangements to be struck between hydrogenerators and other users and for environmental flows. To expand on these issues, this chapter discusses the competing economic and environmental objectives for Australia's major hydrogenerators and the resultant tensions and practices such imperatives have been imposing on the agricultural sector and vice versa.

The chapter itself is arranged into six additional parts. The following section details hydroelectricity reform and the role of grid interconnection. This is followed by a description of the national distribution of the sector. Details of the operation of the Snowy and Tasmanian hydro businesses are provided in section 3, whereas the intersectoral relations between hydroelectricity, irrigation, and the environment are explored in sections 4 and 5, respectively. Some brief concluding remarks are offered in the final section.

Reform of Australia's Hydroelectricity and the Role of Grid Interconnection

Electricity in Australia was once organized on the basis of vertically integrated state-owned enterprises, which largely enjoyed state-based monopolies in generation, transmission, and distribution. The nation possesses only limited hydroelectric resources, thanks to limited water flows and somewhat flat topography. Ironically, however, hydroelectricity played a significant role in the emergence of a national electricity grid and then a national electricity market.

Of the Australian states, only Tasmania possessed the hydrology and topography for hydroelectric development, and the sector thrived there. The development of water power became a de facto state development policy, for much of the development of hydro potential in Tasmania occurred with the express purpose of developing not just water power, but the state's natural resources (principally minerals and forests; Read 1986; Kellow 1996). The physical separation of Tasmania from the Australian mainland by Bass Strait meant that the mainland states could not avail themselves of the advantage of Tasmania hydro potential, which would have integrated well into their thermal systems and could have provided valuable peaking capacity.

The possibility of establishing an interconnection between Tasmania and the mainland was considered by a Commonwealth inquiry (the Zeidler Inquiry) in the early 1980s. It reported in the context of the contentious debate over hydroelectric development in southwest Tasmania, and the prospect of importing electricity was hopelessly politicized by this context, with the hydro proponents seeing interconnection as a threat, and opponents seeing it as a possible alternative. It was not until the late 1990s that Tasmania became sufficiently interested in interconnection for it to become a reality.

This interest was spurred by the development of a national market into which Tasmania could sell its hydro output at a premium as peak electricity. This market had been made possible by interconnection of the eastern mainland states, which slowly ended indigenous approaches to electricity. The initial move had come with the Snowy Mountains scheme, which involved not just the diversion of the waters of the Snowy into the Murray-Darling Basin for irrigation, but the construction of the Snowy Mountains Hydroelectric Scheme as a means of providing peak electricity to Victoria and New South Wales, the systems of which became interconnected by virtue of being connected to the Snowy. Then, the extension of the high voltage transmission line to an aluminum smelter at Portland in the far west of Victoria facilitated the interconnection of that state with South Australia, and the small gap between New South Wales and Queensland was closed in line with a recommendation of the Zeidler Inquiry, as much in anticipation of a national market as any other reason.

This physical interconnection was then accompanied by the adoption of National Competition Policy, which ranged more broadly across the economy than just electricity (and included water), but required states to open up their electricity markets to competition.

In 1998 the electricity industry restructure was formalized by the creation of a national electricity market which draws together the infrastructure of the states of New South Wales, Australian Capital Territory, Victoria, Queensland, and South Australia, with Tasmania participating in 2006 following the commissioning of the 360-kilometer Bass Strait power cable known as *Basslink*. The reforms have disaggregated publicly-owned electricity utilities into corporatized (or privatized in the case of Victoria) generation, transmission and distribution, and retail businesses, with contestability of suppliers down to the level of individual consumers. The broad aims of the restructure have been to encourage "competition and efficiency in the production and consumption of electricity" (ABARE 2004, *31–32*).

With customers, large and small, now having the option to choose their energy retailer and generators competing on the basis of price and opportunities to deliver, the operating environment is vastly different for the electricity industry. For instance, corporatization, that has legislated a requisite for economic responsibility and financial returns to government shareholders, substantially changes the economic imperatives for electricity generators, as does interconnection. The fast start capability of hydroelectricity offers hydrogenerators opportunities that are not available to thermal generators. As well as entering into short and long term contracts with several retailers in different states or trading on a daily basis on the spot market they can also contract their stand-by capacity (see Marsh 1980). This can mean that they are paid without generating, as hydro capacity (able to be dispatched virtually instantly) can serve as both spinning and stationary reserve, thus improving the economics of otherwise predominantly thermal systems. Clearly, for hydrogenerators to operate under any circumstances, the security of access to water is essential, but it is now even more critical as a result of the market environment.

The National Distribution of Hydroelectricity

Table 8-1 sets out where in Australia hydropower was being generated as at 2002. It indicates that there are a number of hydroelectricity power stations outside the Snowy Mountains scheme and Tasmania.

Putting to one side for a moment the major large-scale schemes, a number of small-scale hydropower schemes are represented in Table 8-1. Some are unconnected to the transmission system but linked into local distribution systems (embedded generation) or not linked to these systems at all (nongrid generation; ESAA 2003). Many utilize water flows, from which potential energy would otherwise be lost. For example, the Drop hydropower station in New South Wales was commissioned in 2002 on Australia's largest irrigation channel, the Mulwala canal. It generates a mere 2MW (Pacific Hydro Limited

Table 8-1. *Capacity (in megawatts) of Hydropower Generation Assets in Australia*

	NSW	VIC	QLD	SA	WA	TAS	NT	ACT	Snowy Scheme	Australia
Hydro[1]	25	453	132	0	2	2,276	0	0	3,006	5,894
Pump Storage	240	0	500	0	0	0	0	0	750	1,490
Embedded and nongrid generation	183	93	19	0	30	0	0	1	0	326

Source: Electricity Supply Association of Australia (2003): *30–31*.

[1] Small hydrogenerators that are nonscheduled in the national market are excluded.

2005). Small-scale projects such as this illustrate the flexibility of the hydro-power generation technology.

The number of small-scale hydropower stations has increased since 2001 (ORER 2005) when the federal government's Mandatory Renewable Energy Target (MRET) was introduced. In effect this scheme subsidizes the generation cost of renewable energy with the creation of tradable renewable energy certificates (RECs). The scheme applies to new generation projects designated as renewable by the federal government. Hence, whereas RECs cannot derive from hydro schemes existing prior to the commencement of MRET, they can be created from increases in generation output over a specified baseline arising from efficiency and management improvements. The scheme enables renewable energy sources to compete economically with nonrenewable sources as it has created a market for renewable energy by mandating energy retailers to purchase a specified number of RECs from renewable energy generators (Australian Government 2005). Although the figures are not substantial for very small hydro schemes, these examples illustrate the role the MRET has played not only for the energy sector but also the water sector to maximize the utility of water resources and increase the economic viability of irrigation schemes.

Harnessing flows from irrigation schemes to generate hydropower is not new. For instance, construction of the Hume Dam 16 kilometers upstream of Albury-Wodonga in New South Wales began in 1919. It was built on the upper River Murray to capture water in high-flow periods for use for irrigation in low-flow periods. In 1957 a power station was built to generate 50MW of hydropower from irrigation releases (MDBC 2005a). A number of water storages have been built in New South Wales and Victoria for the purpose of irrigation, drought, and flood control (MDBC 2005b). The integration of small-scale hydroelectricity with these irrigation schemes has seen the emergence of a number of such plants with small capacity, operating only when irrigation water is needed, but making a contribution to meeting summer peak demand. There are, for example, such units in New South Wales at Copeton (22.5MW), Burrendong (19MW), Pindari (5.7MW), and Glenbawn (5.5MW). In Victoria, summer irrigation requirements determine the operation of 135MW of capacity at Eildon and 180MW at Dartmouth, although the 240MW Kiewa scheme is operated more flexibly on winter and spring flows. There is also some similar generation from irrigation flows at Wellington Dam in WA (2MW), and 30MW of capacity was commissioned as part of the Ord River Scheme in that state, well after the irrigation scheme was commissioned in 1963.

The Wivenhoe Dam in Queensland provides a rare example of multipurpose electricity development. The dam provides both drinking water for the city of Brisbane and flood protection. In addition, two 250MW turbine generator and pump sets use thermal generation during off-peak periods to create hydropower to meet peak demand. Queensland was the last state to seriously entertain a substantial hydroelectric development with the Tully Millstream proposal in the Atherton Tablelands in the early 1990s. Its two dams, feeding a 600 megawatt power station, would have inundated 4,000 hectares of land, including about 1,400 hectares which was World Heritage listed.

Whereas the Snowy Mountains scheme was also built for the purpose of providing irrigation storage as well as flood and drought control, its dual purpose is to generate hydroelectricity (see Wigmore 1968). With its 16 dams, 7 power stations, a pumping station, 145 kilometers of interconnected transmountain tunnels and 80 kilometers of aqueducts, the Snowy Mountains scheme generates around 4,500 gigawatt hours per annum. This represents 74 percent of mainland Australia's hydroelectricity (Snowy Hydro Limited 2005). The Tasmanian system with its 51 dams and 27 power stations generates around 9,600 gigawatt hours (Hydro Tasmania 2005a). Given that these two schemes constitute the major hydrogeneration assets in Australia this chapter now focuses on the Snowy and Tasmanian schemes.

Operation of the Snowy and Tasmanian Hydroelectricity Businesses

Snowy Hydro

In recent years Snowy Hydro Limited, the corporatized entity that operates the Snowy Mountains scheme and Tasmania's Hydroelectric Corporation, known as Hydro Tasmania, have transformed themselves from being conceived as dam builders to firmly cast their corporate images as renewable energy companies. This transition has been assisted by concerns about the contribution of fossil fuels to climate change and the rise in the profile of renewable energy spawned by the MRET. The ecological, economic, and social impacts of the loss of vast areas and on downstream waterways have been eclipsed by the reincarnation of these generators as part of the environmental solution instead of the problem. Having shed their developmentalist origins, these generators are now portrayed as the principal source of Australia's renewable energy and as taking a lead role in efforts to reduce the impacts of climate change (Hydro Tasmania 2005a; Snowy Hydro Limited 2005). For instance, after the Snowy Water Inquiry the then Chief Executive Officer of Snowy Hydro was reported to have raised the issue of climate change to defend his company's reluctance to relinquish part of its water allocation to return flows to the Snowy River. He stated that a reduction in Snowy hydrogeneration would be made up in the national market by generators using black coal. Hence, a decision to reduce the capacity of the Snowy Mountains scheme would increase greenhouse gas emissions and efforts to rectify one environmental problem could exacerbate another (Hogarth 1998). Indeed, the Snowy Water Agreement, which relies on government funding of $375 million to implement water efficiency projects to return 21 percent of average natural flows to the Snowy River within ten years (Parliament of New South Wales 2000), demonstrates how the hydroelectricity sector has (at least partially) contributed to water policy outcomes.

The Snowy Water Agreement depended crucially upon a stroke of electoral good fortune for the people of the Snowy River Alliance. Years of lobbying had led to the establishment of a New South Wales-Victoria Snowy Water Inquiry.

Indeed, corporatization of the Snowy Mountains Hydroelectric Authority was made conditional upon the completion of this inquiry and provisions for its scope were made in the Snowy Hydro Corporatisation Act 1997. It reported in 1999 and recommended that 15 percent of average natural flows should be returned to the Snowy River. A Victorian election later that year resulted in a hung Parliament, with three independents, including Craig Ingram representing Gippsland East, holding the balance of power. Ingram ousted a candidate representing the National Party, which had been supportive of River Murray irrigation, and ALP leader Steve Bracks struck a deal to return environmental flows to the Snowy and thus won the support of Ingram and was able to form government. Subsequently, in August 2002, after all governments agreed to change the water sharing and electricity arrangements of the Snowy Mountains scheme, with an overall target of returning 21 percent of average natural flows, the first release of water back into the Snowy River occurred with the release of water from Mowamba Dam.

In December 2005 the New South Wales government unilaterally announced that it would sell its 58 percent interest in Snowy Hydro by means of an initial public offer. The Commonwealth government announced in February 2006 that it would sell its 13 percent shareholding, and the Victorian government followed suit a week later, adding its 29 percent stake, thus making possible a full privatization of the generator. The sale would not have affected water flows down the Snowy, Murray, and Murrumbidgee rivers, because agreements to regulate and secure water flows had been introduced at the time Snowy Hydro was placed on a commercial footing in 2002.

These safeguards to water did little to assuage those downstream users and environmentalists, however, with opposition coming from rice growers, irrigation companies, land owners, the Green Party and elements within the governing (at the Commonwealth level) Liberal-National coalition. The National Farmers Federation took the opportunity of privatization to seek additional protections for users and the environment, concerned over the impact of privatization on recent water resource initiatives in the Basin, such as the National Water Initiative. Although environmental flows for the Snowy River are to increase to 21 percent in 2012 (and ultimately to 28 percent at some unspecified date) from improved irrigation efficiency, both irrigators and environmentalists saw dangers in securing this outcome. In particular, there was disquiet that future flows might not occur at the optimal times for irrigation agriculture, or the environment for that matter, given the high value of water for peak electricity generation.

The general issue of privatization of infrastructure assets had become politicized in New South Wales as a result of public disclosure of some aspects of the deal relating to the controversial Sydney cross-city tunnel. The New South Wales upper house, the Legislative Council, established an inquiry into the Snowy sale and the prime minister, John Howard, yielded to these pressures and announced on June 2, 2006 that the Commonwealth was reversing its decision to sell, which forced both New South Wales and Victoria to follow suit, much to the embarrassment of New South Wales Premier Iemma.

Tasmanian Hydro

The Basslink integrated impact assessment process provided some insight into how Hydro Tasmania endeavors to balance its competing responsibilities of delivering profits and environmental flows. In its negotiations on the volume of an environmental flow for the Gordon River, Hydro Tasmania argued that mitigation had to be both "environmentally and economically sustainable" (Locher 2001, *iv, 124, 125, 257, 258*). This mandate aligned environmental concessions with Hydro Tasmania's economic and management imperatives. An environmental flow is the most practical mitigation option for a hydro-generator. Not only is it easily released through a power station, but it can also generate electricity in the process. This opportunity of generating electricity from the environmental flow for the Snowy River will also be available to Snowy Hydro Limited (2005) when the outlet structures at Jindabyne Dam have been commissioned.

Although the transformation of the Snowy and Tasmanian generators from dam builders to renewable energy businesses has been similar in some respects, important differences stem from their foundational charters. As noted, the Snowy Mountains scheme was built to meet two equally important goals—to generate hydroelectricity, and to divert water for irrigation, thereby droughtproofing western rivers, like the Murray and the Murrumbidgee. The Tasmanian system, conversely, was built for the sole purpose of generating hydroelectricity and ultimately expanded for the implementation of hydro industrialization, whereby large-scale manufacturing and industrial enterprises were attracted by cheap electricity to set up in Tasmania (Thompson 1981). Hence, while town and domestic supplies and a number of irrigation schemes have legislated allocations, Tasmania's hydrogenerator has had considerable control over that state's water resources in the past. Such control has remained in place with the replacement of Tasmania's water management acts, including the Water Act 1957 with the Water Management Act 1999. Indeed, political representatives with farming constituents have claimed that Hydro Tasmania's control over water resources was substantially strengthened by the enactment of this Act such that compensation was assured for the hydrogenerator, but this was not necessarily the case for farmers, who would also be financially disadvantaged if their allocations were varied (Parliament of Tasmania 1999).

Hydroelectricity–Irrigation Relations

Commencing on January 1, 2000, the Tasmanian Water Management Act 1999 was developed to comply with the reforms from the Council of Australian Governments and the National Competition Council framework described in earlier chapters. Whereas the Act vests all water rights in the Crown, it grants a special license to Hydro Tasmania which is entitled to use "all surface water" in hydroelectric catchments with the exception of water already allocated to users prior to January 1, 2000, or by way of legislation and binding agreements

(DPIWE 2001, *21*). Hydro Tasmania's entitlement is the largest water allocation in Australia (DPIWE 2005).

With Hydro Tasmania having priority access to so much of Tasmania's water resource and a lack of correspondence between the timing of generation discharges and that required for irrigation, many farmers rely on obtaining licenses to extract and store water that spills from hydro dams. It should be noted that hydro system operations are optimized to avoid spills (Hydro Tasmania 2001, *13*). Indeed, an important component of the business case for Basslink is that connection into the national market allows Hydro Tasmania to dispatch and derive revenue rather than spill water (Parliament of Tasmania 2006. In periods of high rainfall and flooding (usually in winter) spills are unavoidable. Spilled water has a low surety level for irrigators due to its low reliability. Although they are not charged for spilled water, farmers incur the cost of licenses and building on-farm or on-stream storages. Interestingly, in 2001 Hydro Tasmania notified farmers in the state's Meander Valley region that they would be charged $25.00 a megaliter for additional water that was once available free (Prliament of Tasmania 2001, *14–15*). In response to questions in Parliament, Minister Llewellyn confirmed that Hydro Tasmania had subsequently agreed not to enforce the charge during the 2001–2002 season and that further discussions would take place. Since this time Hydro Tasmania has not re-instated the charge (Carson 2005).

The Tasmanian government has actively encouraged farmers to build dams to store winter floodwaters for use in the dry summer months. For instance, it has subsidized the cost of environmental studies; streamlined the dam assessment and approval process; conducted studies to identify future dam opportunities; and provided funding for a joint venture irrigation scheme, the Meander Dam (DPIWE 2005). Demonstrating the extent of diversification that has occurred in the agricultural sector in Tasmania, between 1986 and 2003 there has been a seventeen-fold increase in water taken under license and stored in on-farm dams (DIER 2003). Hence, with most of Tasmania's water resources dedicated to power generation, the building of on-farm dams currently underpins the development of Tasmania's water resources and its agricultural sector with the cost of dam projects being paid substantially by farmers. Importantly, in 2006 the lowest ever winter rainfall was reported for Tasmania's central districts and its north coast. Adding insult to injury, the Bureau of Meteorology warned farmers of extraordinary climatic conditions due to very low probabilities for spring and summer rain (Mounster 2006, *1*). Hence, even if they were planted, crops might not be finished off for market due to an expected lack of rain. A continuation of weather conditions such as these and the associated uncertainty for farmers could precipitate challenges to the water arrangements in Tasmania outlined here.

The Snowy Scheme has a total storage capacity of around 7,000 gigaliters, which is equivalent to twelve Sydney Harbors and captures around 2,830 gigaliters annually (Snowy Hydro Limited 2006). The corporation has rights to collect, divert, store, and use the water resource for electricity generation and is required to release a minimum volume of water to the Murray and Murrumbidgee Rivers

(Snowy Hydro Corporatisation Act 1997). These irrigation releases have averaged 2,750 gigaliters per year (Snowy Hydro Limited 2006). Although Snowy Hydro does not sell water or get involved in its allocation to irrigators it has "assisted Australian irrigators and farmers in meeting their crop allocations during difficult drought conditions" (Snowy Hydro 2006, n.p.). These arrangements allow irrigators to borrow water from future allocations in drought seasons and to repay the water by not drawing full allocations in better seasons (Johnson 2006).

In contrast, in Tasmania, under a specified licensing regime with its regulator, Section 121 of the Water Management Act 1999 allows Hydro Tasmania to transfer its allocation to other users. These are financial arrangements that secure farmers' access to water at varying surety levels and are separate to licenses obtained by farmers to access spilled water. Although it is not specified in the legislation, the price Hydro Tasmania currently charges irrigators is based on the cost of lost generation (DPIWE 2001). Assurances were given by Minister Llewellyn during debate on the introduction of the water management bills that this would be the case (Parliament of Tasmania 1999). On this basis, Hydro Tasmania would not profit from its nonconsumptive use of the water.

Interestingly, Hydro Tasmania's predecessors were never required to pay the state royalties for the water they used for power generation. Although state governments have collected various levies from electric utilities (usually in the form of dividends postcorporatization), they have never collected explicitly the economic rent available from electricity generation by way of royalties (as they do with mineral resources), preferring instead to use cheap electricity as a lure for development. In this way they stood in contrast with the Canadian provincial governments (Zucker and Jenkins 1984). Importantly, the opportunity cost of committing water to storage and generation was thus never signaled to other, competing users.

Given that so much of Tasmania's water resource is used for power generation, the price of water in Tasmania is, and will continue to be, linked to the price of electricity. The value of water for Hydro Tasmania is variable within the hydro system and is influenced by a number of factors. In the first instance, it varies temporally and geographically. When water is abundant and storages are high, usually in winter, its value is relatively low. Additionally, the same volume of water will generate more energy the higher it is in a catchment due to the greater head. Hence, water taken from lower reaches will have a lower value. Second, Tasmania's hydro system is approaching its limit to supply Tasmania's long-term demand (Hydro Tasmania 2001). It is important to note that this limit is due to precipitation inflows and not generation capacity. Specifically, Tasmania's hydro system can generate 2,265MW, yet the peak generation load is currently 1,790MW (Hydro Tasmania 2005a). Basslink, natural gas, gas-fired electricity, and wind power are now relieving pressure on the hydro system and contributing to Tasmania's energy mix. Third, it is expected that Basslink will increase the demand for water as exports occur into the national market. Indeed, Basslink has been cited as a reason for Hydro Tasmania's reluctance to enter into any permanent agreements to transfer part of its allocation to other users (Hydro Tasmania 2001; DPIWE 2001). There have been two exceptions.

In 2003 an agreement was reached between Hydro Tasmania, its regulator, and farmers to permanently transfer around 33,000 megaliters (DPIWE 2005). Also, in 2005 an in principle agreement was reached between Hydro Tasmania and Gunns Limited for the sale at a commercial rate of an annual 26 gigaliters for the operation of the proposed Bell Bay Pulp Mill from 2008 (Hydro Tasmania 2005b). To put these figures into some context, it is worth noting that during 2001-01 around 37,000 gigaliters passed through the turbines of Tasmania's hydro system and around 1,700 gigaliters represents the nation's water consumption in hydroelectricity production (ABS 2004).

Each of the factors that influence Hydro Tasmania's water value represents considerable unknowns for water trading in Tasmania into the future. As Hydro Tasmania's internal demand for water increases so will the water value and the cost for other users seeking a transfer of allocation from the hydro-generator. Although water trading is still in its infancy (DPIWE 2005), the price irrigators pay to Hydro Tasmania will influence the price at which they can on-sell their allocations. It is unknown at this stage whether the commercial rate payable by Gunns Limited will be greater than the cost of lost generation. If it is, while the agreement for the Bell Bay Pulp Mill could be viewed to accord with an objective of the national water reforms—to derive the highest possible value for water with appropriate price signals—irrigators are likely to view it as a potential threat to their access to water from Hydro Tasmania at the lowest possible price.

Hydroelectricity–Environment Relations

Tasmania's Water Management Act 1999 has instigated the development of water management plans to regulate allocations, license water users, and secure base flows for Tasmania's waterways. The process requires an assessment of the quantity of water required by an ecosystem as well as the detrimental effects of taking water and its impact on water quality. Provision in the legislation for compensation arising from changes in allocations, the introduction of environmental flows, for instance, is an important issue for both the agricultural sector and hydrogenerators. Indeed, an argument for the development of Tasmania's controversial Meander Dam is that without it there would be insufficient water for irrigation needs as well as environmental flows (RWSC n.d., 14). Section 111 of the Act grants "paramount surety" to Hydro Tasmania's water allocation. Ecosystem needs share this surety level. However, Section 116 provides that a variation to a special license to give effect to a water management plan is only binding on the special license holder if there is agreement on the issue of compensation. As the legislation stands, then, compensation would be claimable by a special license holder if its allocation was reduced, for instance, to introduce an environmental flow. An agreement between the Tasmanian government and Hydro Tasmania clarifies this potential clash of objectives. This agreement waives Hydro Tasmania's right to compensation for what is described as an "environmental variation" which might be "required to implement a provision of a water management plan" (BJAP 2002, Appendix 18, Attachment 1, clause

1). The test for what can be designated as an "environmental variation" is specific. As well as the case needing to be made by government that an "environmental variation" "must occur" for the objectives of the Act to be met, it needs to be established scientifically that there are no other "reasonable alternatives," that it is "necessary" for the sustainability of an ecosystem; and that the special license holder is the "direct cause of the environmental effect" for which the variation is required [BJAP 2002, Appendix 18, Attachment 1, clause 1.1(a)-(d)]. These criteria give considerable protection to the hydrogenerator's access to the water resource and its revenue potential.

An "environmental variation" only goes so far, however. For example, a major component of Hydro Tasmania's mitigation package for Basslink is its adaptive management and monitoring program for the Gordon River. A special license amendment between the government and Hydro Tasmania provides for variations to be incorporated into its special license. This amended agreement waives Hydro Tasmania's right for compensation in respect of the variations to the license to give effect to the Basslink environmental flow commitments approved at the close of the impact assessment process. However, the waiver "does not extend to any subsequent variations . . . in relation to the adaptive management commitments" (BJAP 2002, Appendix 18, clause 4). Hence, if it is found that Basslink operations cause unforeseen damage on the Gordon River in the future and there are recommendations for a change to operations to restore the ecosystem's sustainability, Hydro Tasmania will be within its legal right to claim compensation for any loss of its allocation. Although Hydro Tasmania would probably view this provision as protecting its financial interests, an enforcement of this limitation on the waiver for compensation would be at odds with the principles of adaptive management. On this basis, it could act as a political and economic barrier to managing the Gordon River ecosystem in an adaptive way in the future.[1]

The Snowy Hydro Corporatisation Act 1997 is slightly more flexible. Sections 25 and 30 allow for an adjustment of the environmental flow for the Snowy River without compensation after a five-year review. Beyond this point, however, Snowy Hydro Limited will be entitled to compensation if part of its resource is to be surrendered to return more than 21 percent of flows to the Snowy River (Parliament of New South Wales 2000). These cases illustrate the challenge for decisionmakers in balancing stakeholder interests and their demands for economic and resource security within the future needs of ecosystems. They also provide an additional example of the difficulties of managing uncertainty and risk, as described by Quiggin in Chapter 5. The legislative triggers for compensation vested in the hydrogenerators could see other water users disadvantaged and environmental objectives compromised if licensing protocols fail or management plans do not adequately account for the long term.

Conclusion

We have seen that with the restructure of the Australian electricity industry the two major hydrogenerators are operating in a substantially different corporate environment. Water reforms and substantial shifts in the agricultural sector have coincided with these changes. Within this context and because the reform legislation does not adequately account for the potential for change over the long term or for adaptive management, it is possible, if not probable, that environmental objectives will founder. Compensation provisions for allocation variations that can be invoked beyond the short term are effectively fixing allocations and environmental flows on the present horizon where there is little or no evidence of their effect. Although these provisions might meet the short-term economic and management needs of the hydrogenerators (and state governments attempting to privatize electricity assets), such is unlikely to be the case for the ecosystems that the CoAG reforms are intended to make sustainable for future generations. Problematically, it could be some time before the full implications of the resource security afforded to the hydrogenerators are known. We should be mindful, though, that the hydrogeneration sector has already provided us with some poignant lessons in Tasmania and on the Snowy River. In particular these lessons illustrate how legislation, decisions, and worldviews that do not sufficiently account for the potential for future change or lend themselves to adaptive management can seriously undermine our capacity to act when things have clearly gone awry. The new environmentally friendly corporate images of the major hydrogenerators should not tempt us into forgetting the lessons that they have given us from the past.

Notes

1. The issue of adaptive management and its incorporation into water policy is given greater attention by Pagan later in this volume.

References

ABARE (Australian Bureau of Agricultural and Resource Economics). 2004. Energy in Australia 2004. http://www.abareconomics.com (accessed August 25, 2005).

ABS (Australian Bureau of Statistics). 2004. *Water Account, Australia, 2001-01.* Cat No. 4610.0. Canberra.

Australian Government. 2005. Mandatory Renewable Energy Target. http://www.greenhouse.gov.au/markets/mret/ (accessed August 25, 2005).

BJAP (Basslink Joint Advisory Panel). 2002. *Basslink proposed interconnector linking the Tasmanian and Victorian electricity grids: Final Panel Report.* Hobart: Resource Planning and Development Commission.

Carson, G. 2005. Personal communication between G. Carson, Hydro Tasmania, and the first author, August 4.

DIER (Department of Infrastructure, Energy, and Resources). 2003. Rural Land Use Trends in Tasmania 2003. http://www.dier.tas.gov.au/forests/rural_land_trend_2003/content/irrigation.html (accessed November 1, 2005).

DPIWE (Department of Primary Industries, Water, and Environment). 2001. Water Development Plan for Tasmania: Final Report, June 2001, Hobart, Tasmanian government. http://www.dpiwe.tas.gov.au (accessed August 14, 2005).

DPIWE (Department of Primary Industries, Water, and Environment). 2005. Report on the Operation of the Water Management Act 1999, Prepared by Water Resources Division of

DPIWE. Hobart, Tasmanian government. http://www.dpiwe.tas.gov.au (accessed July 26, 2005).

ESAA (Electricity Supply Association of Australia Limited). 2003. Electricity Australia 2003.

Hogarth, M. 1998. Inquiry fails to pacify snow water warriors. *Sydney Morning Herald,* October 24, 7.

Hydro Tasmania. 2001. Report on Hydro Tasmania: Background Report, May 2001. Hobart, Hydro Tasmania. http://www.dpiwe.tas.gov.au (accessed August 14, 2005).

Hydro Tasmania. 2005a. *Hydro Tasmania Annual Report 2004/2005.* Hobart: Hydro Tasmania.

Hydro Tasmania. 2005b. In Principle Water Agreement Reached with Gunns, Media Release, July 22, 2005, Hydro Tasmania, Hobart. http://www.hydro.com.au (accessed October 28, 2005).

Johnson, P. 2006. Personal communication between P. Johnson, Snowy Hydro Limited, and the first author, September 11.

Kellow, A. J. 1996. *Transforming power: The politics of electricity planning.* Cambridge: Cambridge University Press.

Marsh, W. D. 1980. *Economics of electric utility power generation.* Oxford: Oxford University Press.

MDBC (Murray-Darling Basin Commission) 2005a. The River Murray System: Hume and Dartmouth Dams Operations Review. http://www.mdbc.gov.au (accessed August 25, 2005).

MDBC (Murray-Darling Basin Commission) 2005b. Electricity Generation. http://www.mdbc.gov.au (accessed August 25, 2005).

Mounster, B. 2006. Horror dry looms. *Tasmanian Country.* September 8, 1–2.

Locher, H. 2001. Summary Report. Basslink Draft Integrated Impact Assessment Statement report prepared for Hydro Tasmania, Hobart.

ORER (Office of the Renewable Energy Regulator). 2005. Australian Renewable Energy Certificate Registry, Australian Government, Canberra. http://www.rec-registry.com/public/stations.main (accessed August 4, 2005).

Pacific Hydro Limited. 2005. The Drop Hydro Project. http://www.pacifichydro.com.au (accessed November 1, 2005).

Parliament of New South Wales. 2000. New South Wales Legislative Assembly Hansard—Snowy Water Agreement. http://www.parliament.nsw.gov.au/prod/parlment/hansart.nsf/v3key/la20001219012 (accessed August 19, 2005).

Parliament of Tasmania. 1999. House of Assembly, Hansard, Wednesday, June 23, 1999 —Part 3—Pages 101–170, Water Management Bill 1999. http://www.hansard.parliament.tas.gov.au/Parl2.htm (accessed August 19, 2005).

Parliament of Tasmania. 2001. House of Assembly, Hansard, Thursday, November 22, 2001—Part 1—Pages 1–37, Water Charges. http://www.hansard.parliament.tas.gov.au (accessed August 18, 2006).

Parliament of Tasmania. 2002. House of Assembly, Hansard, Tuesday, October 22, 2002 —Part 2—Pages 29–87, Water Management Amendment (Transfer of Water Allocations) Bill (No. 2) 2002 (No. 77) Second Reading. http://www.hansard.parliament.tas.gov.au/Parl2.htm (accessed August 19, 2005).

Parliament of Tasmania. 2006. Legislative Council, Government Businesses Scrutiny Committee A, Hydro Tasmania, Tuesday, July 25, 2006. http://www.hansard.parliament.tas.gov.au (accessed July 26, 2006)

Read, P. 1986. *The organisation of electricity supply in Tasmania.* Hobart: University of Tasmania.

RWSC (Rivers and Water Supply Commission). n.d. Development Proposal and Environmental Management Plan: Meander Dam Proposal. http://www.dpiwe.tas.gov.au (accessed November 3, 2005).

Snowy Hydro Limited. 2005. Business Update. http://www.snowyhydro.com.au (accessed November 1, 2005).

Snowy Hydro Limited. 2006. Corporate, Company Background Fact Sheet, http://www.snowyhydro.com.au (accessed August 14, 2006).

Snowy Hydro Corporatisation Act 1997. http://www.legilsation.nsw.gov.au (accessed November 1, 2005).

Thompson, P. 1981. *Power in Tasmania*. Hawthorn: Australian Conservation Foundation.

Water Management Act 1999. http://www.thelaw.tas.gov.au/index.w3p (accessed August 25, 2005).

Wigmore, L. 1968. *The struggle for the Snowy*. Melbourne: Oxford University Press.

Zucker, R. C., and G. P. Jenkins. 1984. *Blue gold: Hydro-electric rent in Canada*. Ottawa: Economic Council of Canada.

CHAPTER 9

Ecological Requirements
Creating a Working River
in the Murray-Darling Basin

Terry Hillman

*A*S NOTED EARLIER IN THIS VOLUME, the Murray-Darling Basin is nationally important because of its contribution to Australia's rural production (over 40 percent). In addition, it holds an iconic status with much of the population. There is widespread acceptance that current signs of degradation in the system augur future decline; economic, sociocultural, and ecological. As a result the Basin has been a focus for research in recent times. Detailed analysis of the ecology of the Murray-Darling would be a gargantuan task and outside the terms of reference of this book. Instead, this chapter is aimed at setting an ecological context within which analysis of the management of the Murray-Darling as a water resource might be set. The task is approached in the following steps:

- *The context.* A superficial look at the Murray-Darling Basin and its recent history, emphasizing an ecological context.
- *The hydrological footprint of human interventions.* A recasting of management of the river in terms of hydrological changes that have significant (negative) outcomes.
- *Ecological response.* Examining some of the main components and processes of the riverine ecosystem, their links to hydrology, and ways in which the ecosystem responds to hydrological change.
- *Future management response.* Exploring areas in which management changes might be considered as a means of supporting ecological input to the triple-bottom-line debate.

Notwithstanding the difficulty of generalizing to other locales, the chapter provides insights into the complexity of incorporating ecological considerations into the reformation of water policy. The chapter also highlights the imperative of dealing with the demands of the ecosystem in a systematic and propitious manner.

An Ecological Perspective on Natural Resource Management

Human beings are part of an ecosystem and, like all other living organisms, both influence and are influenced by it. Although we are able to create great changes in the landscape in which we live, we are not able to manage all the outcomes of those changes or avoid the repercussions arising from the changed ecologies. Continued successful use of natural resources from our environment demands a clear understanding of our relationship with the landscape and its long-term husbandry. The situation infers a two-way relationship. Conceptually, it applies to all relationships between humans and any other organism or ecosystem that they require to harness and this, by definition, includes natural resource management (NRM).

The term *working river* was coined originally by Dedee Woodside (Hillman et al. 2000) to describe aquatic ecosystems harnessed as a natural resource and is used throughout this chapter to identify river ecosystems supporting a water resource.

For the relationship between working rivers and humans exploiting the resource to be successful over an extended period, a form of contract, albeit unilateral, is required that, in effect, sets out the obligations of the participants. The obligation of the working river is plain; the optimum supply of a water resource into the future. In return, the natural resource manager should seek to

- maintain the resource in a productive condition;
- maintain other nonworking attributes, often integral to the system's long-term function; and
- satisfy a broader duty of care.

These might be referred to as NRM obligations.

The first NRM obligation relates to an optimization of production/exploitation but the latter two have broader ramifications involving the wider community. Here researchers such as Geoff Syme (Chapter 15) are exploring, *inter alia*, means through which the community can be engaged in drafting the NRM contracts rather than having them imposed either by government or by the direct beneficiaries of the resource.

To achieve the aim of indefinite optimization of the resource while satisfying the wider community's aesthetic/ethical/philosophical requirements—in other words, to draft the best possible contract—we need better understanding of:

- the ecological processes that determine the condition or "health" of the resource;
- the processes through which the resource can support optimum production into the future; and
- the means to incorporate the requirements of an informed community in drafting the contract (i.e., in developing policy).

This chapter deals only with the first of these. Though not sufficient, this can be seen as essential, however, because without a healthy river, productive, cultural, and social goals are unlikely to be met in the long term. The remainder of this chapter concentrates on the ecosystem supporting the resource in the knowledge that cultural, social, and economic factors are dealt with elsewhere in this book.

The Resource

Although the Murray-Darling Basin covers almost 14 percent of the surface of Australia it is, in terms of volume, a small river system by international standards. A volume equivalent to the Murray's mean *annual* discharge leaves the mouth of the Amazon every seventeen hours on average. This is partly because the Murray-Darling Basin occupies a fairly dry part of the continent but mostly because Australia is a very dry continent (see, for instance, Fletcher and Powell in Chapter 2).

As well as being comparatively small, flows in the Murray-Darling system are highly variable between years and seasonally unpredictable. This has led to the evolution of a flora and fauna geared to cope with environmental extremes and well adapted to unpredictable conditions.[1] Interannual variability is also a problem for irrigated perennial crops. Unregulated, the Murray-Darling system can be expected to cease to flow several times per century as the fledgling irrigation community in the Sunraysia area discovered in 1915 when sandbags were needed to pond what remained of the Murray to supply water for their pump (Hallows and Thompson 1995). As well as being reliable, irrigation water supply must be timely—available during the hot, dry growing season. Management actions to remedy these two hydrological factors plus the removal of a major proportion of the water for human use are the major drivers for ecological change in the Murray-Darling system.

A Timeline

Warren Musgrave (Chapter 3) describes human history in the Murray-Darling Basin during the past century and a half in terms of three phases:

1. Establishment Phase
2. Development Phase
3. Reform Phase

Progress through these phases also matches changes in our relationship with the Murray-Darling ecosystem. In this light:

- The Establishment Phase marked the confiscation of land largely for grazing and grain cropping. The main ecologically significant activity—ignoring the introduction of rabbits, pigs, and goats in the latter part of the nineteenth century—was the clearing of native vegetation in areas of more intensive land-use. Apart from infinitesimal water supply for stock and domestic consumption, the main human use for the river was for transport. Some de-snagging within the technological limitations of the time and the harvesting of riparian forests for fuel were the main ecologically significant activities. Complete domination over nature was probably seen as a laudable objective if not a divine right, but its realization was severely restricted by limited technology. At the end of the nineteenth century the development of large steam-driven centrifugal pumps greatly enhanced the distribution of irrigation water and large stationary engines were employed in land clearing. By the second decade of the twentieth century the first irrigation-induced salinity problems were literally surfacing in some of the irrigation schemes in Mallee areas, resulting in the development of extensive tile drainage networks as part of returned soldier settlement schemes. This might have been seen as the first signs of an ecosystem backlash had the ecological framework for such an interpretation existed.

- The Development Phase continued its predecessor's objective of complete domination of nature as the means to maximizing returns from natural resources. It also continued the belief that future advances in technology would resolve any current difficulties and that the capacity to move and store water, including water gathered from other catchments, presented the only real limit to water resource exploitation. Rapid technological advances, fuelled by two world wars and associated socioeconomic needs for land and production, provided a heady *ambience* for this can-do approach to managing natural resources. The result was rapid expansion of water storage and diversion for irrigation which culminated in the imposition of a cap on water diversion at a point in time when less than 30 percent of the water expected to reach the sea in the absence of diversions did so.

- The Reform Phase, as Musgrave explains, is yet in its infancy and the remnants of its predecessor are still substantial. Faith in the technological fix still haunts our deliberations—the forlorn belief that a free lunch is out there somewhere waiting for us. By and large, current policy development is carried out in the recognition that our natural resources are components of larger ecosystems that determine their nature in space and time. As will be seen in the remainder of this chapter, progress in this regard often is limited less by the willingness to incorporate the ecosystem in our deliberations than by the lack of knowledge needed to do it.

Ecology is an infant and, until very recently, ill-supported science. The lack of knowledge during most of the past century has resulted in signals of ecological damage being interpreted as negative side-effects for which technological responses are required. As a consequence we have tended to address symptoms rather than look to support or restore the appropriate functions in the ecosystem. While symptom relief has its place in both human health and river health, sooner or later treatment of the cause is required. Addressing symptoms is important, especially if it buys time to attack the problems in a systematic way. However, ever increasing expenditure on salt interception in the face of rising salinities, increasing water-treatment to combat the effects of more frequent algal blooms, or periodic fish release to compensate continuing loss of native fish are not solutions and do not address underlying aspects of ecological decline that may, in time, threaten the resource.

The Human Footprint

It is obvious that the massive rural development that has occurred in the Murray-Darling Basin could not have taken place in an ecological vacuum. Flows in the Murray are now managed to maximize the quantity and reliability of water supply primarily for irrigated agriculture in the Basin—that is to maximize the river's value as a resource for humans, at least in the short term. It may be informative to reverse this approach to the issue by assessing the river, in its current state, as an ecological resource. How well does the river in the first decade of the twenty-first century serve the Murray ecosystem?

Although there have been other changes, the primary anthropogenic effect has been hydrological both in terms of the quantity of water available to the ecosystem, and its spatial and temporal pattern. Changes to the quantity of water are self-evident. Ecologically significant changes have occurred as a result not only of diversion of water for human productive activities but also of the structural and managerial actions that have been instituted to support those activities. Discussed later are listed some of the significant changes, their causes, and their probable ecological affects.

Inter-Basin Transfer

Inter-basin transfer is not a common practice in the Murray-Darling Basin although, historically, the prospect is raised in political circles in times of drought. The exception is the Snowy Mountains Scheme (described in detail by Duncan and Kellow in Chapter 8) which diverts water to the Murray and Murrumbidgee rivers from eastward-flowing systems. The overall effects on aquatic ecosystems is somewhat diffused by other management actions, but in the Tumut River, a tributary of the Murrumbidgee, significant physical damage in the form of bank erosion, sediment transport, and riparian water logging result. Recent interstate agreement has seen the amount of transfer from the Snowy system reduced to manage environmental damage—not in the receiving streams of the Murray-Darling Basin but in the Snowy River.

Physical Barriers

Weirs were built early in European settlement as an aid to navigation, and thereby river transport. These and later purpose-built structures also raise water levels that create pools for pumping and/or hydraulic head for gravity diversion. Larger structures, like dams, provide storages through which seasonal pattern and annual variability of water supply are managed. These hydraulic factors will be dealt with later, but the existence of the structures as barriers to movement up and down the river is also ecologically significant. Several important species of native fish migrate over significant distances either seasonally or in response to flow conditions. Failure to migrate will lead, in at least some instances, to failure to recruit to the population and will at least reduce genetic mixing in the populations. The same is true for many aquatic invertebrates. The barriers also preclude downstream movement or at least make it more hazardous. Any significant bodies of still water upstream of the structure create a barrier to downstream drift—a means of dispersal for many invertebrates and the larval stages of some native fish species—and create an unsuitable habitat for many organisms requiring flow, thus creating significant spaces between otherwise contiguous populations. Larger reservoirs also cause suspended material to settle out along with associated nutrients. As a result rivers downstream of these water bodies are sediment starved and therefore erosive.

Depressed Summer Temperatures

During the warmer months, in water bodies of sufficient depth, a phenomenon known as stratification occurs. This involves the development of a temperature and density differential between upper and lower strata of water such that the upper and lower strata cease to mix, isolating the deeper waters from the warming effects of the sun. The bottom stratum loses contact with the atmosphere as well and the biota gradually deplete the water of dissolved oxygen. Where dams release from the bottom of the reservoir water temperatures in the river downstream are likely to be significantly depressed. In a number of cases the effect can be detected for hundreds of kilometers downstream. Take the case of the Mitta Mitta River where data have been collected on water temperatures between 1978 and 2005. These data show that, in the absence of releases from Dartmouth Dam, summer temperatures average in the mid-twenties—quite suitable for native fish breeding. In years in which significant releases from Dartmouth are made to supplement storage in Lake Hume, for example the early 1990s, water temperatures fail to exceed the mid-teens, too cold to support native fish breeding and marginal for a number of insect species.

Almost all aquatic animals in the Murray-Darling Basin are poikilothermic or cold-blooded and their metabolic rate, development, and behavior are determined by water temperature. This includes such critical processes as hatching of eggs, progress through developmental stages in invertebrates, and movement and breeding behavior in fish. Reaches in which temperature is significantly depressed are probably dependant on their tributaries to maintain viable populations.

Inverted Seasonal Flow

The purpose of major storages is to normalize downstream flow for human productive purposes. In effect this means ironing out interannual variability and controlling seasonal variability to synchronize water delivery with crop demand. The latter task is achieved by capturing catchment runoff in winter and early spring for later release during the irrigation season. This results in an inversion of the pattern of seasonal flow in those areas located downstream of major storages but upstream of major irrigation diversions.

The ecological significance of inverted seasonal flow is complex and is often compounded by other factors; water temperature (noted earlier), riparian land management, and short-term flow fluctuations (noted later) are examples.

Unseasonable inundation of river bank and floodplain is the primary concern. Sustained high flow in spring/summer suppresses the establishment and growth of riparian plants that would otherwise stabilize the river banks. The effect is exacerbated by grazing stock, particularly cattle that effect physical damage as well; turbidity, which reduces light penetration for photosynthesis; and the requirement to carry additional flows from interbasin transfer (Hillman 2004). In concert with long periods of constant flow, high summer flows from dam releases pose a substantial threat to bank stability.

Modified Short-Term Flow Variability

Day-to-day variation in water level is a natural characteristic of rivers. Where the degree of management is high this variation is often reduced—not so much as an integral part of managing the resource but usually as a by-product of extended time-steps in regulating flow or, occasionally, as a matter of pride-in-performance on the part of an individual weir operator. Usually, extended periods of constant river height are punctuated by sharp and significant changes, as demand or other management imperatives require. The result is that the total or mean water level change over a given period may not appear to have altered, although the shorter-term pattern may be modified substantially.

Loss of short-term variation suppresses the productivity of the river by reducing the zone through which sunlight can penetrate. As the water level rises and falls over a fairly limited depth range, photosynthetic organisms attached to the substrate are able to utilize sunlight penetrating the water. The variable light climate also encourages diversity in this community—the biofilm (Burns 1997).

Exaggerated bank erosion tends to occur at the river surface where ripples and wind-driven wavelets create an abrasive movement of sediment particles in a very short-term wetting/drying process. Constant water level concentrates this action at one stratum on the bank, creating an erosion notch. Such notches result in bank slumping, particularly if they are at more than one stratum or if they occur below banks subject to occasional water-logging.

Removal of a Flow-Class

Management of the Murray-Darling system changes the frequency distribution of flow levels in three ways:

- The removal of zero and very low flow events;
- The removal of moderate high-flow events (i.e., minor floods) through their capture in storages; and
- Licensed diversion triggered by flows above a prescribed level, for off-river storage and subsequent irrigation.

Had it not been regulated, the Murray would have ceased to flow at least twice in the past century. As explained earlier, low flows in summer/autumn are no longer likely to occur and demands for town supplies and stock and domestic rural supplies ensure that river managers maintain a significant flow at other

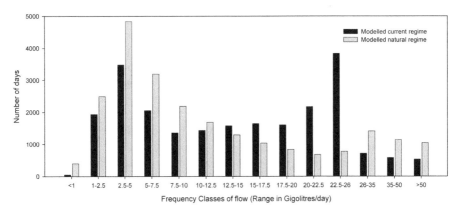

Figure 9-1. *Frequency of Daily Flow Classes at Albury under the Current Management Regime and without River Regulation, 1935–1997*

Source: Data from Murray Darling Basin Commission, courtesy A. Close.

times of the year. Sustained low flows are a threat to bank stability if they result in bank notching.

Figure 9-1 summarizes frequency of flow classes for the period June 1935 to June 1997 in the Murray at Albury, assuming current regulation practices applying over the whole period and assuming no river regulation. The data are derived from BIGMOD, a daily time-step model developed by the Murray-Darling Basin Commission. It is clear that the operation of Hume Dam results in a loss of very low flows, which would have occurred during summer and autumn under natural conditions, and those in excess of 26,000ML/day, flows that result in lateral movement of water onto the floodplain.

In the Darling system there exists a class of water diversion license that permits harvesting of water when flows exceed a specified, high level. This includes the capture of water that overflows to the floodplain. The practice presumably arises from the misconception that such flows are excess to the requirements of the river ecosystem. In fact the result is to deprive the river system of flow events that inundate the floodplain and the large floodplain water bodies typical of the system—and thereby damage the ecosystem components dependant on such events.

Changes in Frequency of Significant Flow Events

Flow events mean more than just water to the ecosystem. They contain information. A number of ecological processes are cued by specific features of the hydrograph, particularly high flow events (HFE). Examples such as fish migration and recruitment and water bird breeding events are discussed further in the following section. The current management regime has generally resulted in changing the frequency of these events. This is important in itself as these events represent the opportunities over time for the organisms involved to proceed through key steps in their life cycle, such as dispersal and reproduction. Any reduction in their frequency represents a significant disbenefit to those organisms. The disbenefit can be amplified if other factors make these increasingly rare events less favorable by increasing other negative factors (e.g., reduced food, limited habitat). A particularly serious threat occurs when the maximum period between these flow events exceeds the life span of the organisms dependant on them. Modeled data show that, had current river management been in place in the Murrumbidgee during the past century, the maximum period between flow events capable of supporting water-bird breeding in its lower reaches would have extended from four to fourteen years—even though the mean frequency of occurrence had only halved (Hillman 2004). As the life-span of these birds is probably seven to ten years the significance of the extended spell is obvious and much more disastrous than information on average changes might imply. In highly variable ecosystems, such as the Murray-Darling Basin, changes to extremes may be critical.

The Ecological Response

Links between human activity and the River Murray ecosystem have been discussed extensively elsewhere (e.g., Walker 1986, Jones et al. 2001, Thoms et al. 1998). The following discussion is limited to factors relating to the hydrological footprint of river management and in particular those components and processes of the river ecosystem that might be better supported by changes to the way in which the resource is managed.

Although the most obvious result of current water use in the Murray system is the removal of water—the volume now flowing to the sea is less than 30 percent of that under natural conditions—most ecological responses are linked to changes in hydrological pattern. Exceptions include the decline of the complex estuarine/freshwater ecosystems of the Coorong (Jensen 1998), associated with the Murray mouth, and the decline of River Redgum in the Chowilla area, near the South Australian/New South Wales/Victorian borders. In the latter case the trees are under stress from the reduction in frequency of high flow events and the recent drought has resulted in extensive damage here and in other parts of the Murray-Darling Basin. There is an additional problem at Chowilla relating to the quantity of water in the system. Hydraulic head provided by the Murray may be no longer strong enough to oppose the saline groundwater flanking the river. Without intervention, Redgums in this area may be threatened regardless of modifications to the pattern of river flows.

In most other cases it is the spatial and temporal pattern of flow that influences the riverine ecosystem. The remainder of this chapter examines the link between flow and ecosystem components and processes.

The Biota

It is not possible to deal with all the organisms that are dependant on the river system—or even to deal with all the major groups. The discussion here is limited to organisms that have some claim to iconic status or play a key role in the river ecosystem.

Fish. There are about forty species of native fish recorded from the Murray-Darling system. An audit of native fish throughout the basin currently under way (The Sustainable Rivers Audit, Murray-Darling Basin Commission) is indicating that, in the catchments examined so far and including the Murray, around 50 percent or less of the native species predicted to be present are being sighted. It is estimated that, over all, native fish are currently at about 10–12 percent of their pre-1750 numbers (Harris and Gehrke 1997) and are influenced directly by a wide variety of factors relating to river management. These factors include in-stream barriers, depressed temperatures, seasonal shift of high flows, and reduced frequency of high flow events. Humphries et al. (1999) demonstrated that different native fish species respond to a variety of flow conditions. Some species, notably Golden Perch, are cued to migrate and breed by flow events, others such as the Murray Cod may attempt to carry out these processes seasonally but are dependant on flow conditions for success. Current research (J. Koehn, Dept. Natural Resources and the Environment, Victoria; pers. com.) indicates that movement by Murray Cod is significantly curtailed where the density of snags (i.e., submerged fallen trees, course woody debris) is reduced.[2]

It is likely that nonflow-related factors such as competition from alien species and, in the past, fishing pressure have a detrimental effect on native fish. However, by limiting successful recruitment and dispersal, *inter alia*, flow management compromises the resilience of native fish populations and thereby their capacity to resist these forces.

Water Birds. Lumping all bird species associated with water under the heading "water birds" is a dangerous oversimplification. The discussion here is limited to those species that require a wetland habitat for successful breeding, though these too may include migratory and nonmigratory birds and encompass a wide variety of aquatic environments. It is assumed here that a mix of such habitats in space and time results in support for a wide variety of water bird species. The provision of this heterogeneity is implicit in the following discussion.

Water bird rookeries occur throughout the Murray-Darling Basin in the form of deflation basins and other large nonpermanent floodplain water bodies. Most, perhaps all, of these have suffered significant decline over the past fifty to one hundred years (Kingsford 2000). In general the hydrological requirements of water birds can be expressed in terms of timing and duration of inundation of these water bodies though, of course, individual species' requirements can be quite complex, involving precedent conditions and the rate of decline of water level amongst other factors. The performance of individual wetlands as bird habitat under changed hydrological conditions is significantly affected by whether the wetland is in the form of a basin (which, once filled, retains water for a period) or not—the latter case requiring sustained high water levels in the host river to maintain inundation. This distinction is critical when considering the rehabilitation of water bird habitats. Wetlands with a water-holding capacity are influenced primarily by the frequency of appropriate high flow events. The value as bird breeding habitat of wetlands that do not retain water is dependant on both the frequency and duration of such events.

If conditions do not remain favorable following the induction of breeding, most water birds will abandon their nests, leaving eggs and chicks to perish. This is a natural response but, as increased modification of flows reduces the opportunity for breeding, such failures become more costly to the water bird communities.

River Redgum. The river redgum, *Eucalyptus camaldulensis*, is the main floodplain tree along the Murray. It grows as a major component of floodplain forests in upper sections of the river gradually thinning to open woodlands as the river enters drier landscapes in the Mallee region. The adult tree (i.e., more than ten years old) develops deep tap roots that reach into the groundwater. Even where this water is quite saline the tree can use it to sustain a low level of metabolism for a number of years. Significant growth, flowering, and seed set are generally dependant on a supply of fresh water to the feeding roots in the upper zone of the soil. Inundation of the floodplain at an appropriate time creates favorable conditions for the dropping of seeds, their dispersal, and germination. Although long-term inundation (i.e., one to two years) is fatal, the river redgum can withstand extended flooding that eliminates potential competitors.

Extended periods without inundation result in tree death through dehydration, the effects of extended exposure to high salt levels, or succumbing to other factors due to a loss of vigor. Flooding in an inappropriate season—summer/autumn—may contribute to adult tree health but is unlikely to support seed dispersal and germination. Flooding at this time when high evaporation rates would otherwise dry out the floodplain may also lead to waterlogging and the death of seedlings. Reduction in the overall frequency and extent of flooding also allows competition from other tree species, notably black box (*Eucalyptus largiflorens*).

Aquatic Plants. Aquatic plants occur in the main channel of the river, backwaters, and floodplain wetlands. They may be totally submerged or emergent and may, particularly in the latter group, continue to survive and/or grow for periods without surface water. Specific adaptation to various permutations of a wetting/drying regime leads to a structuring of the aquatic vegetation community, particularly in wetlands, and supports the diversity of plant communities (Casanova and Brock 2000).

Aquatic plants are important primary producers. They also provide diverse habitat and protection from predators, particularly for invertebrates and small fish (Balcombe and Closs 1996). On river banks aquatic plants provide protection from the erosive effects of flow and, particularly, ripple and wave movement at the surface (Frankenberg 1997).

Seasonal inversion and constant flow levels both have a negative effect on the diversity and extent of aquatic plant communities in rivers, particularly in conjunction with high levels of turbidity. Plant diversity and extent in floodplain wetlands is also dependent on the pattern of wetting and drying. In addition to the hydrological regime, aquatic plants are highly susceptible to land management practices, particularly grazing.

The Floodplains

By convention a floodplain is the land flanking a river, which is inundated by a flow with a return frequency of 100 years or less. In fact a strict definition is not necessary and floodplains are usually apparent from their alluvial soil, specialized biota, and close association with a river channel. There is no floodplain without a river, and the river and floodplain interact to create a single intradependant ecosystem.

Floodplains are populated by organisms that are flood dependant (i.e., they need floods as part of their life strategy) or flood tolerant (i.e., do not need floods but are capable of surviving them). In both cases floods take out susceptible competitors. Spatial and temporal patterns of inundation are the main drivers for floodplain diversity and productivity. The juxtaposition of the various floodplain plant communities into a mosaic of vegetation types results from the interaction between the environmental preferences of the plant species involved and the hydrological regime. Floodplains typically contain a variety of habitats including distributary channels, backwaters, billabongs (both episodic and permanent), wet meadows, and a variety of plant associations, and these support diverse animal communities. The high level of biodiversity is thus the direct result of hydraulic heterogeneity, and changes to the flooding regime constitute a threat.

Barmah Forest provides an example of such changes. Reduction of large floods and seasonal inversion of flows have resulted in increased duration of summer flooding in low-lying areas and a reduced frequency of inundation of higher parts of the floodplain. The result is a severe reduction in the area of wet meadows (i.e., Moira Grass plains) with temporary wetland communities encroaching on the wetter parts of their former distribution and, because higher flooding is less frequent and of shorter duration, redgums are colonizing the drier extremes. The result is that, during the past seventy years, the area of Moira Grass plain in the Barmah Forest has shrunk from over 40 percent to about 5 percent. The bird and animal assemblage associated with this habitat have probably followed a similar pattern.

The term *billabong* is of Australian Aboriginal origin and refers specifically to floodplain wetlands. Billabongs are extremely productive ecosystems (Boon et al. 1990) and support dense and diverse invertebrate communities not found in the river's main channel (Hillman 1986). As well as contributing significantly to biodiversity, these communities provide an important food source to water birds (Kingsford 1998) and, in the form of emerging adult insects, to bats and bush birds. Parkinson (1996) demonstrated a significant increase in numbers and diversity of bush birds in the vicinity of billabongs.

Whether previously dry or not, billabongs respond to inundation with a rapid and substantial increase in productivity. Current research indicates that wetting of previously dry areas produces immediate release of nutrients and dissolved carbon (Baldwin and Mitchell 2000) and sets in motion the following sequence of events:

- Day 1–3: Maximum levels of dissolved and fine particulate carbon present in the water column.
- Day 2–5: Maximum concentration of bacteria in the water coming from the sediment and reproduction in the water column.
- Day 7–14: Zooplankton hatch from drought resistant eggs and reproduce in the water column to increase densities rapidly.
- Day 10 onwards: Macroinvertebrate number and diversity increase (originating from resistant forms in sediment, relict populations in residual water—if any—and eggs from dispersing flying adults).

Wetting sediment also stimulates germination of aquatic plants from a seed bank stored in the sediment or from propigules carried by incoming water, and flooding kills invasive terrestrials.

Floodplains are dependant on out-of-channel flows, and their ecology is responsive to all aspects of such events including timing, size, duration, and frequency. Responses to anthropogenic changes to river hydrology depend on both the nature of the change and the biotic component involved. Because of the complex interactions that characterize floodplain ecosystems, adaptation to changed hydraulics can involve multiple knock-on effects that are often expressed as reductions in biodiversity and modification to ecosystem processes and which may take some years to be fully expressed.

In much of the Murray-Darling Basin, floodplains have been substantially altered by land management practices such as forestry and grazing. Hydrological change independent of river flow management, such as the construction of levees and draining of wetlands, has also been widespread. In terms of river management these factors modify the floodplain's capacity to cope with changes to the flow regime or to respond fully to restoration of flow events.

Ecosystem Processes

As well as affecting individual organisms, changes to river hydrology alter the relationships between those organisms and also the functions they perform as part of a living ecosystem. These changes are difficult to predict and can result in unpleasant surprises. Not all such processes can be addressed here but two are presented as critical examples.

Primary Production. In ecological terms primary production is the term describing the process by which inorganic carbon is captured and turned into organic carbon—the basic form of biological energy supporting metabolic processes and the "currency" of all trophic transactions in the ecosystem. At the level of this discussion primary production is equivalent to photosynthesis, and primary producers can be divided into four groups:

- Phytoplankton: Pelagic algae suspended in the water column of the river or wetlands.

- Biofilm: Photosynthesising organisms (e.g., algae and bacteria) forming part of the complex organic mat that covers underwater surfaces.
- Aquatic plants.
- Terrestrial plants that produce leaves and other material that finds its way to the river or wetlands, contributing carbon to the aquatic ecosystem.

Each of these primary producer groups supports part of the aquatic food-web and therefore their relative representation tends to determine the relative composition of the biotic community at that time. They also tend to be favored, or disfavored, by particular hydraulic conditions. Phytoplankton are favored by deep and preferably stable conditions. They tend to dominate production during high in-channel flow, in weir pools and reservoirs, and in deep wetlands. Unlike phytoplankton, biofilm organisms are static and can photosynthesize only when sunlight reaches them. They are disadvantaged by long periods of high flow and by constant water levels, particularly if the water is turbid. Their relative contribution is also dependant on architectural factors such as snags and aquatic plants. Aquatic plants are also disadvantaged by extended periods of constant flow and by seasonal inversions which result in consistently high water levels during the growing season. Changing water levels in the river or wetlands promotes species diversity and spatial distribution of aquatic plants (Casanova and Brock 2000).

Input of organic material from terrestrial plants occurs either by direct fall into the stream or wetland or through sweeping in litter from the floodplain during floods. The latter, in terms of both production and delivery, is dependant on floodplain hydraulics. Organic material from litter—particularly eucalypt leaves—differs from other forms in that it is resistant to rapid processing by consumer organisms. Periodic inputs of floodplain litter to the river can be likened to stocking the freezer as distinct from the rapid turnover of fresh produce from the other sources. In naturally unpredictable ecosystems, such as the Murray-Darling, this is probably an important distinction.

Connectivity. Connectivity refers to the natural connectedness between physical components of the river ecosystem that permits physical and biological transactions between them. Longitudinal connectivity was already discussed earlier. Many species of invertebrate and some larval stages of fish are known to drift downstream as a means of dispersal or of finding appropriate habitats. This process is inhibited by physical barriers such as weirs or the bodies of still water that bank up behind them. Likewise, several native fish species including Murray cod and golden perch are known to migrate considerable distances upstream, possibly in association with reproduction. This is also disrupted by physical barriers such as weirs. It is ironic that structures built to improve connectivity for paddle steamers had such a profoundly opposite affect on the native biota.

Lateral connectivity describes connecting the river either to floodplain wetlands or to the wider floodplain. The event is characterized by the movement of water out into the floodplain followed by the movement back to the river. As a rule of thumb, when such a connection occurs the floodplain receives water from the river and the river receives biotic resources from the floodplain.

The type and quantity of resources received by the river depend, *inter alia*, on the nature of the hydrological event that creates the connection—particularly its size and duration, but also other factors such as the length of time since the precedent event.

As indicated earlier in this chapter, addition of water to a billabong sets off a chain of events and the nature of material returned to the river will depend on the duration of connection—after one day the return water will contain quantities of dissolved organic material and bacteria; after twelve days it will carry large numbers of zooplankton, which may provide food for larval fish. The significance of such inputs back to the river has not been established scientifically. It is possible that they may recharge backwaters which are important nursery zones for larval fish (King 2003).

The significance of litter input was also discussed earlier but to be effective, such input requires a substantial flow. These events are less frequent than before in the upper parts of the Murray-Darling Basin and becoming increasingly rare in the lower reaches, below the points at which most of the irrigation water is diverted. Current research is examining in-stream productivity, a formidable task in large rivers. Results are not yet available but questions regarding the capacity of these reaches to support significant food webs, including large fish, are yet to be answered.

Optimizing Management

Much of the subject matter of this book deals with means of sharing water between the ecosystem and *direct* human use. As the resource is in limited supply for either purpose, optimum use requires efficiency based on an understanding of the link between the resource and the desired outcome. For rural production these links are well understood, as are potential efficiency measures and their costs. Equivalent knowledge is not available for the ecosystem. As ecological research progresses, however, it is becoming possible to link ecological disbenefits with aspects of the way in which the resource is managed and used for human purposes. Redressing some of these threats may require a shift in the share of water, but others may benefit from changes solely to the way in which the resource is managed. Table 9-1 attempts to evaluate links between management-based changes to the Murray-Darling and key components and processes of the ecosystem as discussed in this chapter. The number of asterisks indicates the severity and the likely ecological effect under current circumstances. Importantly, a number of these can be ameliorated by changes other than to the quantity of water supplied.

Physical barriers, such as locks and weirs, inhibit longitudinal movement of fish and the downstream dispersion of invertebrate and fish larvae. The bodies of water backed up behind them exacerbate their effect on drift, reduce short-term variability in water level, and create inappropriate lateral connection with associated floodplain areas. The Murray-Darling Basin Commission is currently designing and constructing fish ladders for all regulating structures on

Table 9-1. *Some Links between Anthropogenic Changes to Flow and Likely Responses from Components and Processes of the River Ecosystem*

| | Ecosystem Response | | | | | | | | | |
| | Biota | | | | | Primary Production | | | | Connectivity |
Flow-modifying Factor	Fish	Water-birds	River redgum	Aquatic Plants	Floodplain	Phyto-plankton	Biofilm	Aquatic Plants	Floodplain Vegetation	Connectivity
Inter-basin transfer			*	***	*	**	**	***	*	**
Physical barriers	*****			**	**	*****	***	**	**	*****
Depressed summer temperatures	****		*	*		**		*		
Seasonal flow inversion	**	**	****	****	****	**	**	****	****	***
Reduced short-term variability				*****		**	*****	*****		*
Loss of flow class	***	*****	****	***	*****	**	****	***	*****	****
Changed frequency of flow events	****	*****	***	**	*****	*	**	**	*****	*****

KEY:

Can be redressed by modification of existing structures

Effect can be ameliorated in some cases by new structures

Might be improved (for flows less than channel capacity) by changes to management without affecting supply

Note: The strength of the link between each ecosystem response and flow-modifying factor is represented by the number of asterisks (more asterisks = stronger link and/or greater response).

the Murray between the barrages at the river mouth and Hume Dam, nearly 1,500km upstream. This should not be seen as a panacea, addressing only the upstream movement of a subgroup of native fish and leaving other aspects of longitudinal connectivity untouched, but it does provide for some important, iconic species permitting migration and relieving predation pressure on artificially crowded populations downstream of weirs.

In some, but not all, cases the effect of weir pools on lateral conductivity could be reduced by regulating structures. These could be used to provide more appropriate water regimes for connection with floodplain systems and for floodplain vegetation.

Summer water temperatures are depressed when downstream flows are drawn from the bottom of reservoirs. This happens at most if not all of the major storages in the Murray-Darling Basin, and could be redressed, obviously, by drawing water from other levels. This requires some form of multi-level offtake to provide the capacity to draw as much as possible of the water from the storages. Retrofitting such equipment is expensive—particularly as it offers no benefit to the major productive users of the resource. Stronger motivation might be forthcoming in the future if other restorative measures in rivers downstream of storages make depressed temperatures an obvious limiting factor, or where recreational values provide a financial incentive.

Although they affect only a proportion of the rivers of the Murray-Darling Basin, seasonally inverted flow regimes are very damaging, particularly when combined with interbasin transfers, as is the case in the Tumut River. The phenomenon is an inescapable result of management for irrigation supply and is unlikely to be substantially changed. Its quite serious effects on floodplain systems can be ameliorated by the use of flow regulators where conditions allow. This method has been used very successfully at several sites along the Murray (by the New South Wales Wetlands Working Group) fitting regulators to billabongs and larger floodplain lakes (e.g., Moira Lake) and restoring ecological function—with a net gain of water to the river. Management using regulators could become an important means of supporting bird breeding events in the future.

Loss of short-term variability is an unfortunate spin-off of flow management, either in getting the maximum supply downstream or because weir settings are maintained constant for long periods. There is some pressure to do this to simplify the setting of pump off-takes and to maintain constant head for gravity-feed systems. In the latter cases some compromise level of weir manipulation should be developed, particularly where aquatic plant growth and diversity and bank stability are under threat. Where maximum supply capacity is the driver, tolerance of minor flooding should be developed. The Pratt Water Group in a recent study of the Murrumbidgee (Pratt Water Pty Ltd 2004) suggested the possibility of using off-stream en route storages to provide (*inter alia*) variable summer flows in the Tumut River.

The previously mentioned threats have in common the fact that the steps canvassed as possible remedies have little or no effect on the supply of water for irrigation. The remainder, however, relate to aspects of river flow as it is currently managed for irrigation supply. Above all else the acquisition and use of water for ecological sustenance needs to be strategic. The following points might be seen as fundamental:

- Retiring water from productive use (either through reduced allocations or reliability of supply) is a last resort and should be linked to increased use efficiency where possible.
- The application of water for ecological purposes must be based on the best available ecological knowledge.
- The use of water for ecological outcomes should be based on an adaptive management framework so that the activity automatically highlights knowledge gaps and supports means of addressing them.
- The application of ecological allocations will need to be strategic and flexible and matched to an understanding of current ecological needs. There will never be enough water to sustain the ecosystem if it is applied according to rigid and uncritical rules.

In the absence of a broad knowledge base, a first step in satisfying ecological needs is to examine the current hydrological regime in comparison with the situation modeled in the absence of current developments and use. This exercise together with our current understanding of aquatic ecosystems at least affords the opportunity to identify likely areas of concern.

A recent study of the Murrumbidgee River showed that high flow events with the potential to support bird breeding occurred in the lower reaches of the river about half as frequently as they would have done under natural flow conditions (Hillman 2004). Superficially this could have been argued as an acceptable trade-off against productive use of the water. However, spell analysis of modeled data for 100 years showed that the longest period between these flow events under natural conditions was four years whereas, under current conditions of management and diversion, that maximum had extended to fourteen years—bad news for birds that live for seven to ten years! A similar situation with regard to lateral connectivity events was also demonstrated. The message is clear. Although on average the current level of hydrological modification might be tolerable, it is likely to be unacceptably damaging in extreme circumstances. In other words management protocols need to be developed that incorporate a sliding scale of urgency for specific, ecologically significant flow events so that any reserve of water gained for ecological purposes can be used for maximum benefit. A trial use of environmental reserve, piggybacked on ambient flows, to sustain water levels in Barmah-Millewa forest in 2001 resulted in successful recruitment for a number of water bird species and demonstrated the power of this approach. Current research at the Australian Bureau of Agricultural and Resource Economics (Beare et al. 2005) is aimed at developing novel means of ensuring that water normally allocated for irrigation use can be made available for this type of strategic response under appropriate circumstances. Ultimately, maximizing human and ecological benefit in water use will require equivalent water use efficiency on both sides of the equation. For this we need (*inter alia*):

- Significantly increased knowledge of the links between hydrology and the riverine ecosystem;
- Flexible and responsive mechanisms for managing 'environmental reserves' in major storages; and
- Additional novel means of reserving and accessing water for the environment other than through bulk storage and release (e.g., Beare et al. 2005).

To be content with developing means of coping with current conditions would be to totally misunderstand our situation in a period of unprecedented anthropogenic change. The best available predictions suggest that rainfall along the Great Dividing Range in south eastern New South Wales and northeastern Victoria—the source of most of the Murray-Darling flow—will drop by about 10 percent in the next twenty to thirty years. At the same time mean temperatures are predicted to rise by two degrees, resulting in significantly increased evaporation. The outcome is as obvious as it is unpalatable. Adjustment on a scale much greater than that which is currently proving contentious is likely to be unavoidable.

Notes

1. It might be argued that one negative aspect of river regulation is that it has deprived native species of the competitive advantage that such adaptations would otherwise have bestowed in the face of competition from introduced species.

2. Murray cod generally use snags as bases. Their movement along a river is significantly facilitated if the distance between bases is short.

References

Balcombe S. R., and G. P. Closs. 1996. Macrophyte use by fish assemblages in the littoral zone of a Murray River billabong. Paper presented at the Australian Society for Limnology 35th Congress, Berri, South Australia.

Baldwin, D. S., and A. M. Mitchell. 2000. The effect of drying and re-flooding on the sediment and soil nutrient dynamics of lowland river-floodplain systems: A synthesis. *Regulated Rivers: Resource Management* 16:457–467.

Beare, S., R. Hinde, T. Hillman, A. Heaney, and I. Salbe. 2005. Meeting environmental outcomes: A planning framework. OECD Workshop on Agriculture and Water: Sustainability, Markets, and Policies. Adelaide and Barmera, 14–18 November 2005.

Boon, P. I., J. Frankenberg, T. Hillman, R. Oliver, and R. Shiel. 1990. Billabongs. In *The Murray*, edited by N. Mackay and D. Eastburn, 182–199. Canberra: Murray Darling Basin Commission.

Burns, A. 1997. *The role of disturbance in the ecology of biofilms in the River Murray, South Australia*. PhD diss., Adelaide University.

Casanova, M. T., and M. A. Brock. 2000. How do depth, duration and frequency of flooding influence the establishment of wetland plant communities? *Plant Ecology* 47:237–250.

Frankenberg, J. 1997. *Guidelines for growing phragmites for erosion control*. Albury, NSW: Murray-Darling Basin Commission.

Hallows, P. J., and D. G. Thompson 1995. *The history of irrigation in Australia*. Mildura: Australian National Committee on Irrigation and Drainage.

Harris. J., and P. Gerhke. 1997. *Fish and rivers in stress: The NSW rivers survey.* Cronulla, NSW: NSW Fisheries and the CRC for Freshwater Ecology.

Hillman, T. J. 1986. Billabongs. In *Limnology in Australia,* edited by P. De Deckker and W. D. Williams, 457–470. Melbourne: CSIRO.

Hillman, T. J. 2004. *Murrumbidgee Valley Ecological Assessment.* Supporting document for The Report of the Murrumbidgee Valley Water Efficiency Feasibility Project. Campbellfield, Victoria: Pratt Water Pty Ltd.

Hillman, T. J., J. D. Koehn, D. S. Mitchell, D. Thompson, J. D. Sobels, and D. Woodside. 2000. *The Murrumbidgee: Assessing the health of a "working river."* Report to Irrigated Agribusiness Taskforce and the Department of Land and Water Conservation.

Humphries, P., J. D. Koehn, and A. J. King. 1999. Fishes, flows and floodplains: Links between Murray-Darling freshwater fish and their environment. *Environmental Biology of Fishes* 56:129–151.

Jensen, A. 1998. Rehabilitation of the River Murray, Australia: Identifying Causes of Degradation and Options for Bringing the Environment into the Management Equation. In *Rehabilitation of rivers: Principles and implementation,* edited by L. C. de Waal, A. R. C. Large, and P. M. Wade. New York: Wiley.

Jones, G., A. Arthington, T. Hillman, R. Kingsford, T. McMahon, K. Walker, J. Whittington, and S. Cartwright. 2001. *Independent Report of the Expert Reference Panel on Environmental Flows and Water Quality Requirements for the River Murray.* Report to the MDBC.

King, A. 2003. Identification and characterisation of larval fish-nursery habitats in floodplain rivers of the Murray-Darling Basin. PhD diss., La Trobe University.

Kingsford, R. 2000. Ecological impacts of dams, water diversions and river management on floodplain wetlands in Australia. *Australian Journal of Ecology* 25:109–127.

Kingsford, R. T. 1998. *Management of wetlands for waterbirds.* In *Wetlands in a dry land,* edited by W. D. Williams, 111–122. Canberra: Environment Australia, Biodiversity Group.

Parkinson, A. 1996. Macrohabitat use by birds on the Ovens River floodplain. PhD diss., Monash University.

Pratt Water Pty Ltd. 2004. *The business of saving water: The report of the Murrumbidgee Valley Water Efficiency Feasibility Project.* Campbellfield, Victoria: Pratt Water Pty Ltd.

Thoms, M. C., P. Suter, J. Roberts, J. Koehn, G. Jones, T. Hillman, and A. Close. 1998. *River Murray Scientific Panel on Environmental Flows: River Murray from Dartmouth to Wellington and the Lower Darling.* Canberra: Murray Darling Basin Commission.

Walker, K. F. 1986. The Murray-Darling River System. In *The ecology of river systems,* edited by B. R. Davies and K. F. Walker, 631–694. Dordrecht: Dr. W. Junk.

CHAPTER 10

Urban Water Management

Geoff Edwards

A RELIABLE SUPPLY OF HIGH QUALITY WATER at a reasonable cost is of utmost importance for households and for many industrial and commercial firms. Threats to the safety of water supplies are taken very seriously by water users and by governments, a striking illustration being the establishment of a dedicated body responsible for catchment management and raw water quality in response to the discovery of the parasites *Cryptosporidium* and *Giardia* in Sydney's drinking water in 1998. Concern over Australia's urban water supplies in recent times, however, has focused less on the quality of urban water, which is largely taken for granted by water users, than on the balance between demand and available supplies. The word *crisis* has been used to describe the situation facing the urban water systems in Perth, Brisbane, Sydney, and Melbourne as well as many smaller cities and towns. The recent concerns have resulted mainly from drought-induced below-average runoff into urban water catchments. Projections of increasing consumption with population increase and rising incomes, combined with predicted reduced water availability because of climate change and increased environmental allocations, add a longer-term concern to the shorter-term problems from a run of low rainfall years. The Victorian government contends that, in the absence of substantial reductions in per capita water use, Melbourne's consumption could be approaching its supply limits before 2020 (Victorian Government 2004), whereas Perth, Sydney, and other large centers are concerned about how to balance water consumption and available supplies much earlier.

The aim of this chapter is to encourage a re-think about the functioning of the water economies in Australia's major cities. Among the questions that arise in doing so are: Why is water treated differently from other consumer goods that are essential for city-dwellers, such as fuel and fresh food? Are there sound economic reasons, such as the monopoly situation of the urban water authorities or the need to ensure conservation of water, for the heavy regulation of the price and use of urban water—compared with the primary reliance on competitive markets for most other consumer items? Why is integration of the regional and urban markets for labor and capital proclaimed as enhancing economic efficiency, whereas for water " . . . current government policies continue to separate or balkanise the urban and rural water markets" (Freebairn 2005, 4)? Do governments put reductions in urban water consumption ahead of economic efficiency in water use? Would there be a substantial impact on low-income households if greater use was made of price to ration water? When customers have told governments and government-owned water businesses that they are supportive of restrictions on water use in order to save scarce water, have they had good information on the options for saving water and for adding to supplies? Are the roles of the water pricing regulators and their relationships with the governments to whom they report well defined?

Most of these questions are addressed in this chapter. The focus is on water for residential customers, which represented 57 percent to 70 percent of water supplied by the water authorities (excluding environmental flows) in Australia's mainland state capital cities in 2004–2005 (Water Services Association of Australia 2005). For water supplied to commercial and industrial customers, the information available on charging, water restrictions, and other matters is more limited than for the residential sector. Another aspect of urban water economies that receives little attention here (for space reasons) is disposal of wastewater.

The chapter is organized into four further sections. First, some physical and political economy realities that characterize the urban water economies are recognized. Although rainfall and topography largely determine how much water can be harvested easily for an urban population, using water efficiently in a city requires that it be allocated to those users for whom and to those uses in which it has the highest marginal social value. How closely that is approached in urban water economies is a function of political economy: governments can choose in their urban water policies to emphasize economic efficiency or they can instead give priority to social and political objectives. Second, different approaches to charging for water are considered. In doing this, attention is paid both to cost recovery approaches and to managing the demand for water; the former is taken very seriously by state governments in pricing water, whereas the latter is pursued more by nonprice measures than by pricing policy. Third, three options for improving the balance between water demand and available supplies for Australia's major cities are considered. The options are: more efficient management of existing water supplies; accessing new water sources; and integrating urban and rural water systems. The message emerging from the brief concluding comment is that water shortages will not be a problem for Australia's largest cities in the foreseeable future unless there is a shortage of good water policy. That would be a problem of political economy rather than physical constraints on water availability.

Characteristics of Australia's Urban Water Systems

An appreciation of key physical and political economy features of Australia's urban water economies helps in understanding the how and why of their operation.

Physical Realities

Water storages are large relative to annual use, reflecting variability in annual system inflows, and limited integration of urban water systems with the much larger rural water systems. Variations in annual rainfall impact on water in storage through demand-side effects as well as through the more obvious supply-side effects. In wet years more water runs into dams and less is taken out for watering lawns and gardens than in dry years. To illustrate, in March 1998 (a dry summer) Sydney Water Corporation supplied 2,400 million liters of water per day, compared with 1,700 million liters per day in March 1997 (Sydney Water Corporation, 1999).

Sydney and Melbourne have water storage capacity equal to approximately five and four years' use of water, respectively, at 2004–2005 consumption rates. The Sydney Water Corporation (2005) says that "[t]o ensure a reliable water supply for Sydney, it is necessary to store a greater amount of water per person than is necessary for any other comparable part of the world." The high ratio of water storage to annual consumption viewed as essential for Australian cities is usually explained by the variability in annual rainfall, or, more precisely, in runoff into storages. However, another reason is the limited integration of urban and rural water systems. With Canberra obtaining its water from the Murrumbidgee River basin, Adelaide taking water from the Murray, and sharing of water in the Thomson River dam which links the Melbourne and Gippsland water systems, there are physical linkages between some urban and rural systems. However, the institutional arrangements to allow economic integration, with win–win trade between the rural and urban water economies, generally do not exist.

One of the physical realities of urban water supplies is that the high level of variability in annual runoff not only increases the optimal water storage capacity—other things being given—but it also makes it harder to identify changes in the long-term average annual rainfall and runoff. Sydney's average rainfall in the five years 2000–2004 was 20 percent lower than the average for 1961–1990. Wahlquist (2005) reports Blair Trewin from the National Climate Centre saying "Sydney's weather, though dry, fits into long-term patterns, with a similar dry period early last century." Although it may be convenient for the managers of water authorities and for governments to blame climate change for recent water shortages, meteorologists are cautious about accepting that average long-term rainfall patterns have changed.

Political Economy Realities

The facts of the economics and politics of urban water are less immutable though no less complex than the physical realities. Within the constraint of being accountable to their electorates, state and territory governments determine the main features—and also much of the functional detail—of urban water systems. However, as noted earlier in this book, states and territories have been increasingly subject to the discipline of Commonwealth-induced reforms thanks largely to the Council of Australian Governments (CoAG) agreement in 1994 and the more recent National Water Initiative (NWI).

The state and territory governments have ultimate ownership rights of surface and ground water in their jurisdictions and, regardless of the trend toward privatization in other sectors, urban water remains largely in the hands of government. The privatization movement which has characterized government-owned utilities and other commercial activities such as banking and transport in Australia in the last quarter-century has been absent in urban water. State and territory governments have seen political advantage in retaining ownership of the key urban water utilities. However, most of the utilities have been transformed from boards to corporations, increasing their commercial focus, but still needing to accommodate the noncommercial, political objectives of their government owners.

Urban water systems have natural monopoly attributes, and state-established monopolies are responsible for sourcing water, storing it, and making it available to retailers that typically have local monopolies in distribution. A list of the main institutions and the functions undertaken by each, in the urban water systems of Australia's capital cities is shown in Table 10-1. In some cities, some water service activities are contracted to private firms. For instance, the management of SA Water's metropolitan water and sewerage operations applied this

Table 10-1. *Institutions Undertaking Functions of Urban Water Systems in Australia's Capital Cities, 2004–2005*

	Catchment management	Bulk water	Reticulation	Wastewater treatment
Sydney	Sydney Catchment Authority	Sydney Catchment Authority	*Sydney Water*	*Sydney Water*
Melbourne	*Melbourne Water*	*Melbourne Water*	*City West Water South East Water Yarra Valley Water*	*Melbourne Water*
Brisbane	*South East Queensland Water*	*South East Queensland Water*	Brisbane City Council	Brisbane City Council
Adelaide	Adelaide and Mt Lofty NRM Board	*SA Water*	*SA Water*	*SA Water*
Perth	*Water Corporation*	*Water Corporation*	*Water Corporation*	*Water Corporation*
Hobart	Hobart Water	Hobart Water	Hobart Water	City Councils
Canberra	*ACTEW Corporation*	*ACTEW Corporation*	*ACTEW Corporation*	*ACTEW Corporation*
Darwin	*Power and Water*	*Power and Water*	*Power and Water*	*Power and Water*

Note: Italicized entries are government-owned corporations.

Source: Productivity Commission (2005a) and websites of water businesses.

public/private mode of governance for fifteen years from 1996 (Productivity Commission 2005a, *196*).

Under the CoAG/NCP reform framework, the main reforms required for urban water were in institutional arrangements and in pricing (Productivity Commission 2005b, *26*). Details of these reforms were provided earlier in this book (see McKay, Chapter 4). In reviewing the progress of different jurisdictions against the initial reform framework, the Productivity Commission (2005b, *27*) found that urban water reform " . . . for the most part, [was] well advanced, with the widespread introduction of a consumption-based component in charges to help discourage overuse and implementation of financial cost recovery by service providers to ensure better signals for new investment." Similarly, the Commission found that "[v]arious institutional reforms have also been implemented to increase the commercial disciplines on, and the accountability of, those entities delivering water and sewerage services, with most jurisdictions having corporatized their urban water authorities."

Notwithstanding this progress, the NWI agreed to by CoAG in June 2004 commits governments to continue reform that will allow better management within a market framework of water resources in rural and urban uses. Agreed outcomes of the NWI for urban water reform are:

• provide healthy, safe and reliable water supplies;
• increase water use efficiency in domestic and commercial settings;
• encourage the use and recycling of wastewater where cost effective;
• facilitate water trading between and within the urban and rural sectors;
• encourage innovation in water supply sourcing, treatment, storage, and discharge; and
• achieve improved pricing for *metropolitan* water (consistent with paragraph 66.i. to 66.iv.)[1] (CoAG 2004).

A National Water Commission has been established to help drive the NWI process, and over $2 billion has been allocated toward funding of activities under its auspices.

Some current realities of urban water systems sit ill with the emphasis on efficiency and markets in the statement of outcomes for the urban NWI reforms. Politicization, social engineering, and illiberal regulation are more conspicuous in the working of those systems than consumer sovereignty, competition, and markets. Prescriptive regulations, inefficient block pricing structures, and usually costly recycling of waste water are chosen for moving water consumption and water supplies toward balance. Desalination plants are part of urban water planning in Perth, Gold Coast, and Sydney and are under consideration elsewhere. Water from desalination may be a reliable backstop, but will often be expensive, even if zero cost is assigned to the large emissions of greenhouse gases. Meanwhile, the economically obvious step of allowing and facilitating win-win trade in water between the rural and urban sectors, subject to social cost-benefit analysis of the necessary investment in infrastructure to link the two sectors, is neglected.

Most states and territories have introduced independent regulators to determine prices for urban water. Because most of the urban water suppliers are government trading enterprises (GTEs), independent regulators provide a way of avoiding the conflicts of interest that could arise with governments effectively regulating their own businesses. All state and territory acts dealing with the regulation of water services have a number of conflicting objectives—for example, ensuring sustainable water supplies, looking after the interests of water consumers—and provide little guidance on the weights to be attached to each. It is therefore left to regulators to make judgments about those weights. In reality, governments can and do influence the working of their independent regulators in a variety of ways, including requiring ministerial approval of price determinations, specifying how assets should be valued in determining the costs of providing water and rates of return on assets, and influencing staffing of the regulators.

Charging for and Rationing Water

In this section the price and nonprice measures used in urban water economies to allocate water and to recoup costs are examined. It is helpful to remember that the financial, economic, and political objectives of urban water systems include:

- recovering an appropriate level of costs of the water system from customers;
- providing a return to governments on investments in water infrastructure;
- allocating available water resources efficiently; and
- providing incentives for efficient augmenting of water supplies, including raw water and recycled water.

It will be suggested that the water regimes of Australia's largest cities target cost recovery and government revenue objectives more effectively than the objectives of efficient water allocation and water system development.

A summary of water charging arrangements for the mainland capitals and some other cities is shown in Table 10-2. The cities are listed in order of average water use per residence in the five years to 2003–2004. In all cities except Gold Coast (where there was a negligible increase), water use per residence was lower in the subsequent two years (average for 2004-2005 and 2005-2006) than the average for the previous five years, the falls being largest in Canberra, 50 kiloliters (16.6 percent), and Sydney 39 kiloliters (15.8 percent). Water restrictions, price increases, and moral suasion all played a part in the reductions in consumption in 2004–2005, as did the relatively high rainfall in some cities in that 2004–2005.

The water access charge per residence varies considerably. After Newcastle, Sydney and Melbourne had the lowest access charges. Comparison of volumetric charges for different cities is complicated by the inclining block tariffs (IBTs) used in all cities other than Newcastle, Gold Coast, and Darwin. Under IBTs the price per kilolitre is low for an initial amount of water and " . . . target more discretionary use of water by increasing the price per kilolitre once a base allowance has been exceeded" (WSAA 2005, 13). In Adelaide and Sydney there are two blocks; Melbourne, Brisbane, and Canberra have three, whereas Perth

Table 10-2. *Water Use Per Residence, Water Charges, and Use Charges as Share of Water Bill: Selected Cities*

	Newcastle	Melbourne	Gold Coast	Sydney	Brisbane	Adelaide	Perth	Canberra	Darwin
Residential Water Use (kL)									
Av 1999–2000 – 2003–2004	210	220	221	246	256	261	299	301	468
Av 2004–2005 – 2005–2006	201	190	222	207	224	234	272	251	444
Charges at July 1, 2006									
Access ($p.a.)	35	61	120	64	113	148	155	75	103
Water Use ($kL)	1.14, 1.10	0.82, 0.96, 1.42	1.16	1.26, 1.63	0.91, 0.94, 1.20	0.47, 1.09	0.49, 0.73, 0.95, 1.27, 1.59	0.66, 1.29, 1.74	0.69
Water use charges as share of water bill									
100kL p.a. (%)	77	57	49	66	45	24	24	47	39
250kL p.a. (%)	89	78	71	83	67	57	49	78	63
400kL p.a. (%)	93	87	79	89	78	71	63	87	72

Note: Sewerage charges excluded. Movement to the next water price block in each city is as follows: Newcastle 1000kL per year; Melbourne 40 and 80kL per quarter; Sydney 100kL per quarter, Brisbane 50 and 75kL per quarter; Adelaide 31.25kL per quarter; Canberra 25 and 75kL per quarter; Perth 37.5, 87.5, 137.5, and 237.5 per quarter. Water use charges for Melbourne are for its largest water retailer, Yarra Valley Water Ltd. Charges by other retailers are very similar to those for Yarra Valley Water. Water use charge as share of water bill calculated by author, using charges current as of July 1, 2006, and assuming, for cities with block tariffs that are defined on a per quarter basis, that water consumption in each quarter is equal.

Source: Water Services Association of Australia and National Water Commission Facts (2007) for residential water use; water authority websites for water charges.

has five. Prices for block-one water have been set at much higher levels in Sydney, Brisbane, and Melbourne than in Adelaide, Perth, and Canberra. Higher block-one prices have the efficiency advantage of giving small households a stronger incentive to save water, but they are perceived by some as inequitable because they increase the cost of households' base allowance of essential water. Newcastle and Gold Coast, which do not have IBTs, have water prices above the block-two prices for Melbourne and Brisbane, and approaching the block-three price for the latter city. Newcastle is the only city with a *decreasing* block tariff. However, because the second (top) block cuts in at the very high level of 1,000 kiloliters per year, and because the block-two price is only marginally lower than the block-one price, the effects of Newcastle's volumetric charging policy are little different from those of a uniform water price. Although comparison of marginal social costs of water supply would be necessary for a full assessment of the water charging policies in different cities, it is likely that such an assessment would find as follows: Sydney (high block-one price) and Canberra (low block-one price for a small amount of water and a relatively high block-two price) score better on economic efficiency than Perth (complexity of five blocks, with low prices and high volumes for blocks one to three) and Melbourne (moderate prices and relatively large volumes for blocks one and two).

Although economic efficiency points to ensuring that households' access charge not be lower than the fixed costs of servicing them, reducing access charges and increasing the price per kilolitre of water used is an option for reducing water use in a revenue-neutral way, albeit increasing between-year revenue variability for the water authorities and perhaps for the governments to which they pay social dividends. For typical households using 250 kiloliters of water in 2006–2007, the use-based component of their water bills was highest (in excess of 80 percent) in Newcastle and Sydney, and lowest (49 percent) in Perth (Table 10-2). For households using a low 100 kiloliters of water in 2006–2007, the volumetric charge was less than half the total water bill and the access charge more than half, in all cities other than Newcastle and Sydney. For households using a high 400 kiloliters, the volumetric charge was greater than 70 percent of the water account in all cities except Perth, and it exceeded 80 percent in Newcastle, Sydney, Melbourne, and Canberra. It appears that Newcastle, Sydney, and Melbourne, with the lowest access charges, have the least scope for further water saving by substituting higher water use charges for part of their access charges.

There are different views on the appropriate role for considerations other than economic efficiency in determining water charges, and more is said on this matter later. However, the bottom section of Table 10-2 shows information that some view as relevant in assessing equity in water charging arrangements. The information can be seen as suggesting that Canberra, Newcastle, and Melbourne rate relatively well on equity while Gold Coast, Brisbane, and Perth (and arguably also Adelaide and Darwin) rate relatively badly. This interpretation draws on three factors shown in Table 10-2. First, adding the access and volumetric charges, the average cost to households using a low 100 kiloliters per year (a base allowance of essential water, say) is relatively low (less than $1.50 per kiloliter) in Canberra, Newcastle, and Melbourne and relatively high (more than $2 per kiloliter) in Gold Coast, Brisbane, and Perth. Second, although in no city does the average cost per kiloliter of water rise—as might be seen as desirable for equity—as water consumption increases from 100 kiloliters to a moderate 250 kiloliters, the falls in average cost are smaller, in both absolute and percentage terms, in Canberra, Newcastle, and Melbourne than in all other cities. Third, with an increase from 250 kiloliters a year to a high 400 kiloliters, the average cost per kiloliter of water rises in Canberra and Melbourne—perceived as appropriate for equity—and falls slightly in Newcastle. In all other cities the average cost decreases more, absolutely and as a percentage, than in Newcastle.

Notwithstanding their prevalence in pricing urban water in Australia, IBTs do not rate well on the criterion of economic efficiency. (See, for example, Boland and Whittington 2000; Sibly 2006; Edwards 2006.) One way to explain this is to invoke the *law of one price*. Competitive forces work to generate one, efficient, price for identical goods and services. If regulators keep prices different for different customers, they create incentives to engage in trade to arbitrage the price differences. Although for most urban households the net gains from inter-household trade in water would probably not be worth the trouble of arranging hoses or water-carts and measuring the volume of water transferred between properties, some will find it worthwhile. The price incentive for interhousehold water trade is strongest in Perth, where in 2006–2007 the block-one price was $0.49 per kilolitre and the block-five price was more than three times this amount at $1.59 per kiloliter.

The economist's view on trade that brings prices for an identical item closer to a common level is that it enhances efficiency and community well-being: trade is a good thing. If, however, governments determine that different prices are socially appropriate for households using different amounts of water, can interhousehold trade which removes those price differences be seen as increasing efficiency and community well-being? If one views the government legitimately performing a social engineering role in urban water management, one might answer no, and perhaps support another regulation prohibiting trade in water between households in order to avoid eroding the price differences resulting from the original regulation. This acceptance of a big brother approach to water management in the cities is not easily reconciled with the underlying premise of National Competition Policy. Supported by all Australian governments, this premise is that Australians gain from competition through markets.

Another way of expressing the inefficiency of IBTs is that, because they result in different marginal water prices for different households, the marginal private value of water is not equal across all households. Is it plausible that the price differences between households correspond to different *social* marginal valuations, identified astutely by governments? To answer in the affirmative would be to support adjusting upward the marginal private valuation of water use within block-one by a wealthy individual living alone to equate the marginal value of water within higher blocks by larger, poorer households. That sort of adjustment of private valuations to obtain social valuations is not consistent with the egalitarian instinct of Australians.

The introduction of IBTs in Australian cities has been motivated more by notions of fairness than by efficiency considerations. The rationale for this is open to question given the very small percentage of household expenditure that is accounted for by water. Data available from the ABS shows that in the mainland states and territories spending on water (together with sewerage) in 2003–2004 ranged from less than fifty cents a day in Queensland to $1.21 a day in South Australia for the average household. This was a little over 1 percent of total daily spending on goods and services in South Australia, and significantly less than 1 percent of spending for all other jurisdictions. For households in the lowest income quintile, presumably the households of greatest concern to governments in developing the social components of water pricing policies, spending on water averaged less than $0.80 per day in all jurisdictions other than the ACT, and less than $0.50 per day in Tasmania, Queensland, and New South Wales. Only in Victoria and the ACT was spending on water by the lowest-income quintile greater than 1 percent of total spending on goods and services.

The very small share of water in household spending raises questions about the equity case for departing from a simple, efficient policy of one price for all water. Governments do not contrive to provide households with a base allowance of other essential items such as food, fuel, and clothing—all of which rank much larger in household budgets than water—at less than commercial prices. Governments are able to do so for water because of state ownership of the water resource and of the monopoly status of water businesses, and they find it attractive to do so because they perceive political gains from populist urban water policies. A watery irony in this situation is that many households in the lower income quintiles, as well as the upper ones, probably spend more on bottled water at thousands of dollars per kilolitre than on piped water and sewerage. However, in the event that governments *are* concerned about the effect of a higher water price on low income households, it would be possible to target assistance to those households—and to do so without holding down their water price and likely increasing their water use.

Another question concerning social philosophy can be raised here. Given that water is effectively owned by the government on behalf of the people, do governments focus too much on people as consumers of water and not enough on people as owners of the water resource? If economic rents from the extraction and sale of urban water were returned, in equal shares, to each resource owner, most people would have an ownership interest that was larger than their consumer interest, making high water prices attractive to them. That is, a majority might then support monopoly pricing of water—which present regulation aims to prevent.

Although the available supplies of fresh fruit and vegetables are rationed by price, other measures are more important in the case of urban water. At mid-2006, Newcastle and Darwin were the only cities shown in Table 10-2 that did not have water restrictions, with the restrictions being most severe in the Gold Coast. An incomplete list of the measures that have been used in Australian cities in recent years would include:

- Prohibitions on the use of water for some purposes—for example, watering lawns, washing paved surfaces and cars, and operating slippery slides;
- Restrictions on the days or hours that lawns and gardens can be watered;
- Restrictions on the means by which water is used for some purposes—for example, watering of gardens using soaker hoses or hand-held hoses fitted with a trigger nozzle, but not using sprinklers;
- Offsets system for filling swimming pools and spas—that is, approval to fill the pools is conditional on households saving an equivalent amount of water elsewhere;
- Mandated water-saving fixtures for the house and garden—for example, dual-flush toilets, low-flow shower heads, water tanks, and soil-moisture or rainfall sensors with automatic watering systems;
- Subsidies on water-saving fixtures, appliances and gardening items—for example, shower heads, water tanks, washing machines, dishwashers, garden mulch.

In several cities "permanent low level restrictions" (WSAA 2005, 15) that include measures from all but the last of the six categories listed above have been introduced.

Two related characteristics of the restrictions used by state governments to reduce water use are of particular interest. First, although banning or constraining particular uses of water, the restrictions do not *require* households to reduce the amount of water they use. The effect of the restrictions is to reduce water use by households in aggregate, but individual households are free to increase their water use providing they abide by the restrictions. Second, the restrictions are prescriptive. They require households to avoid some uses of water and they place conditions—time of use of garden watering systems, use of hand-held hoses only if fitted with trigger nozzles—on other uses. First principles indicate that prescriptive regulation as practiced in the urban water sector has two major disadvantages: it does not directly target the variable of fundamental interest—total use of water—and it denies households the freedom to choose how to make any reductions in water use required of them.[2] Recent estimates of the annual cost of water restrictions per household in the ACT are $18 to $24 for Stage 1 restrictions, $198 to $360 for stage 3 restrictions, and $369 to $769 for Stage 5 restrictions (Pearce 2005).

A nonprescriptive form of regulation to reduce a city's water use would be to impose quotas (or entitlements) on each customer's water use. The quotas might be set X percent below the customer's average use in the preceding three years, for example. How would the quotas be enforced? Cutting off a household's water supply if it exceeded its quota would be unacceptable on social and public health grounds. An alternative would be to charge a penalty price for over-quota water. That would effectively mean the quota scheme became an IBT, with the block-one price applying to a household's water quota. Another, more market-oriented approach would be to allow trade between households—and preferably also other urban water users—in water quotas. Households using water in excess of their quotas would have to buy the difference in the quotas market. This approach has in-principle appeal on efficiency grounds and also practical difficulties, but deserves imaginative investigation (see Crase and Dollery 2006).

An objection to water quotas is that they are difficult—and costly—to implement, with major problems in allocating quotas fairly. If two households had the same average water use in the last three years, but one household has just experienced an increase in the number of its members while the other has just become smaller, is it fair that they be given the same water quota? And what water quota would be allocated to new households? These difficulties would need to be taken seriously in considering the introduction of household water quotas. For perspective, however, it must be noted that the prescriptive regulations that would be replaced by the household quotas have significant costs, including, in the view of some, unfairness, especially in the treatment of larger households. It is possible that the current rules determining the *cans, can'ts, and conditional cans* in water use are easier for the water authorities than a household water quota scheme, especially if a soft attitude is taken to enforcing them, whereas the non-prescriptive quota would be preferred by most households.[3]

Governments and their water businesses have also used education and moral suasion in their efforts to reduce water use. Many people have probably found helpful the information provided recently on water bills allowing comparison of water use by their own household with use by other households having similar characteristics. Unlike much other useful information on water, such as the difference in water requirements for European and native gardens, comparative water use details for different categories of households requires information that only the water-supplying businesses have. The mandated labeling of washing machines and dishwashers provides information to potential buyers on the water-efficiency of the appliances. Many householders, however, regard other factors, such as ease of unloading and length of the washing cycle, as more important than the amount of water used, and see little incentive to economize on water at present water prices.

The moral suasion card has been led often by governments trying to reduce urban water use. Appeals have been made to water users to make do with less to ensure adequate water for their children and grandchildren. Victoria's Premier Steve Bracks authorized newspaper advertisements saying Melbournians needed to take fewer baths and to reduce their average time in the shower from six minutes to four minutes. For some, "Changing behavior requires constant reinforcement; for example, enclosures with all water bills reminding people to take shorter showers, to clean their teeth with the tap off. . . ." (Millis 2004, 57–58). T. Dwyer (2005) and Henstock (2005), amongst others, have pointed to the role of guilt in inducing people to reduce their water use. Whereas governments and the water authorities they own report a high level of popular support for their campaigns to reduce water use, not all observers have been impressed. Watson (2005, 7) writes "the every last drop . . . rhetoric is being used to beat urban consumers over the head with crass advertising campaigns about water saving."

Policies for urban water in Australia's major cities reflect the priority assigned to the objective of recovering the costs of the water systems over the objective of using water and associated labor and capital resources efficiently.

The consequences of public water enterprises running financial deficits are of greater concern to governments than economic inefficiency resulting from multiple water prices, and from inappropriate relationships between water price and its marginal social cost.

The short-run marginal cost (SRMC) can be defined as the cost of meeting a unit increase in demand with existing water system capacity. When consumption is low relative to capacity, as in Darwin or Hobart, SRMC is low, and a low water price is economically efficient.[4] What is appropriate when consumption is high relative to capacity? Then, the scarcity of water relative to demand needs to be reflected in its price. Ng expresses this as follows: "When consumption is limited by supply capacity, the opportunity cost of water is not the supplier's short-run marginal cost (typically low), but the marginal demand price" (Ng 1987, 21). Victoria's Essential Services Commission interprets the situation of high consumption relative to capacity differently from Ng, though the logical implications for water pricing appear to be similar. " . . . in times of water shortage" the SRMC includes "the costs of customers not being able to consume as much water as they otherwise would" (ESC 2005, 5). The widespread and continuing use of nonprice water restrictions over several years means that urban water prices have consistently not reflected the opportunity cost of water.

Long-run marginal cost (LRMC) incorporates all the costs of adding to water supply by enhancing system capacity. Pricing water at long-run marginal costs is implicit in the National Competition Policy/National Water Initiative, and it is the approach supported by state government regulatory agencies. With lower-cost ways of adding to supplies being taken up first, LRMC will increase with successive additions to capacity. Does the consistent use of LRMC pricing mean that the relative price of water will rise over time as system capacity is expanded to meet higher demand with increases in population and incomes? Perhaps the most that can be said is "yes, other things being equal." However, with economic integration of rural and urban water systems—considered in the next section— the marginal cost of raw urban water would be the price of buying water from the rural sector. That price would rise and fall with changes in the system-wide availability of water and in the profitability of irrigation enterprises. Potentially, it could, and for economic efficiency *should*, also vary with the demand for water for environmental flows.

Neither the SRMC nor LRMC approaches to water pricing lend any credence to increasing block tariffs. A fundamental requirement for economic efficiency in an urban water system is that all water users face identical equimarginal incentives to save water in all uses, and water suppliers—including potential suppliers—have identical marginal incentives to examine all possible ways of augmenting water resources. Australia's major urban water systems do not meet this desideratum. The package of price and nonprice measures used to allocate available water in the capital cities ensures that householders do not have a common marginal incentive to save water in all possible ways. For water suppliers, the regulatory regimes and political influences make it difficult to properly assess the benefits from extra water (inside water use is unrestricted at often-low prices, although some outside uses are banned and others restricted) and distort comparisons between ways of adding to water supplies (building new dams and facilitating rural-urban water trade are out of political favor whereas water recycling and capture of storm water are in favor).

Future Options for Urban Water

In all mainland capitals, and many other cities and towns, the need for action to achieve a sustainable balance between water use and available water supplies is recognized. Possible ways of achieving this are considered here under three subheadings.

More Efficient Management of Existing Water Resources

There is general consensus amongst state governments and their water authorities that reductions in per capita water use in their cities can be and need to be made. Often, reduced per capita water use is equated with increased efficiency of water use. For economic efficiency, however, the key requirement is that the water available is used where it is most valuable. Only by accepting that the package of multiple prices for water, bans and other restrictions on its use, mandated water-saving items in homes and gardens, and subsidies (at various rates) on a range of appliances, plumbing fixtures, and gardening inputs serves to equate the marginal social value of water across all uses and users can it be said that urban water use is economically efficient. Accepting that would surely be too much!

Notwithstanding that the urban water authorities have been placed on a more commercial basis, it is inevitable with present institutional arrangements that political considerations will play a large part in determining the means by which water is allocated in the cities, and hence the economic efficiency of water use. Nevertheless, economics provides some pointers to reforms that would improve the economic efficiency of urban water use: perhaps political constraints will come to be less of a barrier to increasing efficiency for urban water, as they did in earlier times for reform in other areas—interest rates and industry protection, for example.

For economic efficiency it would be appropriate to move to demand management regimes that rely less on prescriptive measures. One way to do this—the way that is used to ration agricultural and mining commodities—would be to rely on price to limit demand to available supplies. That would require raising price when water is scarce and decreasing it when supplies are abundant.[5] Use of one price for all water would be more efficient than inclining block tariffs (Sibly 2006; Edwards 2006). If IBTs are retained for political reasons, their efficiency costs can be reduced by having no more than two price blocks, and making the block-one water volume sufficiently low that nearly all households are in block-two (Edwards 2006).

The following question may help in thinking about reducing urban water use. If water use needs to be reduced, why not *require* households to reduce their use—not the situation now—and allow them to choose *how* to reduce it? That approach would respect the preferences of individuals, rather than imposing the government's values on them, as occurs with the current bans and other restrictions. If the implementation difficulties noted earlier make that approach—involving household water quotas—unattractive to governments, another nonprescriptive way to achieve the same reductions in water use would be to charge a relatively low price for basic water and a substantially higher (penalty) price for additional water. If the volume of basic water were equal for all households, this approach would correspond to an IBT.[6]

A welcome source of continuous, small savings in water is the reduction of losses through leakage and theft. However, keeping water prices down by relying heavily on nonprice water regulations to manage demand weakens the incentive for water authorities to invest in reducing water losses.

Accessing New Water Sources

Traditionally, when urban water systems approached their capacity, a new dam would come on-stream. That approach to increasing water supplies is now out of favor, partly because of the greater emphasis placed on environmental factors.[7] Perth is the only city for which a new dam has been built in the last twenty years (WSAA 2005, 4), though the Queensland Premier, Peter Beattie, announced before being re-elected to office in September 2006, that new dams would be built in southeast Queensland at Traveston on the Mary River and in the Logan River catchment. Subject to proper evaluation, new dams deserve to be considered along with other options for supplementing water supplies and saving water use.

Recycling of wastewater *is* in favor. Some cities have targets for increasing the proportion of wastewater recycled from present low levels. However, recycling has cost disadvantages that limit its economic use. Unless recycled water is purified to the extent that it can be combined with other water in urban system storages, a separate pipe is needed to take the water to users. So-called third-pipe systems are expensive, though less so if designed into new urban developments than if retrofitted to existing residential areas. There are concerns that, encouraged by the lobbying of environmentalists and engineering interests, politically correct water recycling is being undertaken that does not measure up on economic grounds. Victoria is examining a $1.5 billion recycling plan to divert to Melbourne and other cities and to environmental flows 116 gigaliters of fresh water now used each year in coal-fired power stations in Gippsland. That water would be replaced by treated wastewater, reducing by 85 percent the effluent released into the ocean at Gunnamatta on the Mornington Peninsula (Baker 2006).

Rainwater tanks are also in political favor to allow householders to harvest water resources from their own roofs. In most jurisdictions rebates (subsidies) encourage the installation of tanks, whereas a rainwater tank or solar energy unit is mandatory in new houses in Victoria. Householders often see water tanks as a bad idea because of space and aesthetic considerations.[8] Nor are they generally a cost-effective way to save on bought water at present water prices. Yarra Valley Water (2004, 24) reported a study finding that water tanks did not pay for themselves in Melbourne in thirty years. Mandating their installation is hard to reconcile with economic responsibility or individual sovereignty.

A significant development in Australia is the commitment to building water desalination plants in Perth, Gold Coast and, if water in its dams falls to 30 percent, Sydney.[9] Desalination is a very reliable source of water. However, water from desalination plants often costs much more than water from conventional sources. There is a risk that the desalinated water will not be needed, hence proving *extremely* costly, " . . . if we return to a cycle of reliable and consistent rainfall" (WSAA 2005, *26*). Moreover, it is likely that the desalinated water would not be needed—because of the cutting back of demand—if water prices were lifted to levels somewhat lower than the prices that will be required to cover the marginal cost of making fresh water from sea water.

Many questions are unanswered about the integration of desalinated water into urban water systems. These include questions about pricing, risk bearing, and the provision of choice to water users between high-security desalinated (manufactured) water and lower-security natural water. These questions apply also to other new sources of water.

Before Western Australia's state election in February 2005, Liberal opposition leader Colin Barnett committed to building a 3,700 kilometer channel to bring water to Perth from the Kimberley if elected to government. Consistent with a tradition in Australia of visionary water projects being spared the indignity of having to pass an economic test, no cost-benefit analysis was presented for Barnett's proposal, which was strongly criticized by economists. Although the channel idea has been abandoned since the Liberals' defeat in the election, a pipeline from the same source appears not to have been ruled out (WA Water 2005).

The approach to identifying new sources of water for Australian cities and towns often serves the residents of these urban centers poorly. The National Water Commission says it is important " . . . to ensure that all feasible water supply options are on the table; where certain options are ruled out even before evaluation there cannot be transparent debate about the alternatives, and communities may be saddled with less cost effective options" (National Water Commission 2006, *12–13*). In this context, an interesting question was asked by Malcolm Turnbull, the current Parliamentary secretary with carriage of the Commonwealth's water policy initiatives. The question was: Are Queensland and New South Wales examining the option of providing extra water for southeast Queensland from the large rivers in northern New South Wales as an alternative to piping water much further from the Burdekin River in northern Queensland (Turnbull 2006)? At the core of this question is the problem of engendering interjurisdictional cooperation, an issue that features strongly in the history of Australian water policy (see Musgrave, Chapter 3).

Integrating Urban and Rural Water Systems

This option for enhancing the security of water supplies could logically have been included under the previous heading. It is addressed separately because of its potential importance and its neglect in public debate about the development of urban water systems.

Key hydrological background is that agriculture uses 67 percent of the water extracted in Australia, with 9 percent used in households and 7 percent in industry (ABS 2004). Discussing the situation in countries such as Australia where agriculture is the dominant user of water, Rogers et al. (2002, 9) write: "A small transfer of agricultural water can meet all the demands in urban and industrial sectors. . . . since the required transfer is small, the required price increment is also small." However, the Victorian government's white paper, *Securing Our Water Future Together* (Victorian Government 2004), for example, did not recognize trade with irrigators on the Thomson or Goulburn River as an option for adding to Melbourne's water supplies.

Others have seen a compelling case for adding rural-urban trade in water to the options for examination in supplementing urban water supplies. WSAA (2005, 21) summarizes as follows: "With the exception of Sydney, all Australian capital cities have the opportunity to access water from agriculture through water markets without the need to build substantial infrastructure. Many regional cities are also in a good position to trade with agricultural users as urban and irrigation systems are often interconnected."[10] The Productivity Commission (2005b, 204) says: "In the Commission's view, if water is to be allocated to its highest value use in the future, the urban and rural water markets will need to become increasingly integrated." The federal Treasurer expressed a similar view: "Australia needs to move to fully integrated national infrastructure markets in gas, electricity and water" (Costello 2005, 16). In 2006 several other studies also drew attention to the economic case for trade in water between the irrigation and urban water sectors (Young et al. 2006; Quiggin 2006; Productivity Commission 2006; Business Council of Australia 2006). However, whereas some irrigator-urban trade in water has occurred and more is mooted, it cannot yet be said that the *idea* of further developing such trade has won general community or political acceptance.[11]

Preliminary results on the contribution of rural-urban water trade to alleviating reductions in water supplies in Adelaide, Melbourne, and Canberra were reported by G. Dwyer et al. (2005). Using the general equilibrium model TERM-Water, Dwyer et al. (2005) examined the effects on water use and on regional gross product of a 10 percent reduction in availability of urban water and, separately, a 10 percent reduction of urban and rural water availability. The focus in this study was on Adelaide, Canberra, and Melbourne, and the main irrigation districts of the southern Murray-Darling Basin and Gippsland. The general equilibrium responses to the reductions in water availability were considered for three trading scenarios: no trade in water between regions, trade between regions in the southern Murray-Darling Basin and between urban and rural regions where there is some connectivity, and full trade. The second scenario allows trade between Melbourne and Gippsland, connected via the Thomson dam, whereas the third would require investment in infrastructure to link Melbourne with the Goulburn River Basin.

The results from this modeling work show that cuts of 10 percent in water use in the three cities become reductions of one percent or less in the presence of full trade. Under partial trade, the fall in Melbourne water use is 3.6 percent, reflecting the restricted capacity for trade in that scenario. For Adelaide and Canberra, however, partial trade allows slightly higher water consumption than full trade, where they face more competition for water. When water supplies fall by 10 percent in the irrigation regions as well as in the urban areas, trade does not diminish the general equilibrium effect on urban consumption as much as when supplies fall only for the cities. Even so, the fall in water use is reduced in each city by more than half under full trade and by slightly more for Adelaide and Canberra and somewhat less for Melbourne under partial trade.

Dwyer et al. (2005) also gave estimates of the effects of reductions in water supplies on regional gross product. Most of the reductions that occurred in gross product for the cities under no trade were eliminated under trade in the case of 10 percent reductions in urban water, and the falls in city gross product in the case of urban and rural water reductions were also much smaller under trade. However, with reductions in urban water, each of the six irrigation regions experienced larger reductions in gross product under rural-urban trade than under no trade. For regions other than the Mallee, that result applied also for reductions in urban and rural water. For the three cities and six irrigation regions combined (mainland southeast Australia) full trade was beneficial as measured by total gross product: the reduction in aggregate gross product was smaller by $184 million in the case of a reduction in urban water supplies and by $113 million for a reduction in urban and rural water.

The finding of Dwyer et al. (2005) that rural regions experience lower gross product with rural-urban water trade helps in understanding the opposition to rural-urban water trade in irrigation areas. This opposition has been expressed at the highest level by Gary Nairn, former Parliamentary secretary to the prime minister: "Proposals to transfer water from one catchment to another should only be pursued when all other viable alternatives have been exhausted . . . If you allow inter-basin transfers then you're picking winners and losers, and metropolitan areas are always going to have a greater capacity to buy water than the agricultural sector" (Hodge and Bachelard 2005, 4). This view appears to be inconsistent with the agreed NWI outcome of "facilitat[ing] water trading between and within the urban and rural sectors." In national competition policy there is a presumption that competition is generally better than its absence. In a similar vein, Watson (2005, 6) argues "the obligation to justify continued rigid separation of irrigation and urban supplies should be the other way around."

The suggestion that the rural labor or capital market should be segregated from the urban market would be ridiculed. That segregation makes no more economic sense for water. Irrigators voluntarily selling water would be winners, like the urban water buyers. Denying those with property rights in water the right to sell water outside the rural sector is a restraint on rights that would not be contemplated for other assets. If there are adjustment issues that arise with the trade of rural water to urban centers they can be addressed directly through existing taxation, social security, and other policies (see Freebairn 2003).

Conclusion

The complex water policy regimes in Australia's capital cities result in several costs that would be absent with the straightforward policy of relying solely on a uniform price for all (drinking quality) water to allocate the resource. Those costs include: the cost to water users of not being able to use water as they would like; the economic efficiency cost that results from not providing the conditions for equating the marginal value of water across all water users; the cost of not supplementing water supplies in the lowest-cost ways; and the cost of developing, publicizing, administering, and enforcing the water regulations. Another possible cost is resentment and erosion of social capital resulting both from people observing others violating the water restrictions and from reliance on community members reporting their recalcitrant neighbors to enforce the restrictions.

Notwithstanding the costs of the regulatory water management regimes, governments report substantial support for them by water users. With governments being responsive to the potential election consequences of urban water policies, it is plausible that a majority of voting water users may prefer packages involving bans, lesser restrictions, multiple prices, and subsidies for water-saving measures than reliance solely on a uniform water price to ration water use. Many may prefer the inconvenience cost of water restrictions to the financial cost if a higher water price were used to reduce water consumption. The increasing block price structure, also, is consistent with populism. It is not surprising that water users would strongly support higher prices for households that use more water than they do. Consumers of food, petrol, and many other items might do so too if asked! For these items, some of which are just as essential to people as water, and larger budget items, contriving to make price an increasing function of the amount purchased is not on the policy agenda. However, not everyone wins under populism, and there may be a substantial minority of people who would prefer to pay higher water prices and have fewer restrictions on water use.

A pertinent question in thinking about the claimed popular support for regulatory approaches to urban water concerns is the adequacy of the information available to householders expressing that support. Have householders been asked if they would prefer approaches that, unlike the present regulations, do not restrict particular uses of water but require reductions in total household water use, leaving the choice of where to save the water to them? Do they understand that the mix of regulations and multiple water prices makes it hard to do proper assessments of investments that add to water supplies? Has the "less water use is better" mentality that pervades the urban water regimes blinded people to the opposing liberal tenet that "our cities can afford to have as much water as they are prepared to pay for" (Turnbull 2006)? Do householders know that options exist for win-win trade between the rural and urban water economies?

Rural-urban water trade that resulted in a relatively small movement of water from the large-volume rural water sector to the small-volume urban sector could remove concerns about water availability in most and perhaps all capital cities for the foreseeable future. As well as enhancing the security of urban water supplies, rural-urban water trade—and the possibility of urban-rural trade—would facilitate the determination of raw water prices on an opportunity cost basis for the urban sector. Rural-urban water trade could also stimulate examination of far-reaching reforms that introduced competition to the sourcing of raw water, the supply of bulk water to retail distributors, and even the retail distribution function. It remains to be seen whether these advantages will continue to be outweighed by the political benefits attendant on the current regulatory regimes in Australia's urban water sector.

Acknowledgments

Helpful comments from Robert Dumsday, Gavan Dwyer, John Freebaim, David Godden, Neil Sturgess, Alistair Watson and Ross Young are gratefully acknowledged.

Notes

1. Paragraph 66 i–iv refers to "continued movement to *upper bound pricing* by 2008," development of efficient pricing policies for recycled water by 2006; and development of cost-effective pricing policies for trade wastes by 2006. Upper-bound pricing means pricing at long-run marginal cost.

2. For a discussion of regulation in Australia, including the costs of "excessively prescriptive" regulation—among which are resource misallocation, compliance costs, and the costs of developing and administering regulation—see Banks (2004).

3. Of 14,442 Melbourne customers reported for breaching water restrictions since stage one restrictions began in November 2002, only three have been penalized (Smith 2006).

4. Hobart is the only capital city where, apart from the area in Brighton Council, metering is not used, necessitating a zero volumetric charge for water (Langford and Piccinin 2004, 76; Sibly pers. comm.). That may well be an economically appropriate pricing policy for Hobart.

5. With the price elasticity of demand for urban water likely in the range -0.2 to -0.5 (OECD 1999), a 20 percent increase in price would reduce water use between 4 percent and 10 percent. Because some decisions having a major impact on water use (for example, the design of gardens, changes in watering systems, installing household recycling systems, and fitting rainwater tanks with connections to toilets) are long-term decisions, water use—like petrol use—will be more responsive to price in the long-run than in the short-run.

6. A conventional quota scheme would set quotas as some proportion of recent household water use for each household, *requiring* each household to reduce its water use or face a penalty. Relying on a two-block IBT with a common amount of block-one water for each household does not require any household to reduce its water use, but households in aggregate *voluntarily* reduce water use.

7. In an insightful critique of urban water policy, T. Dwyer (2006, *14*) writes: "The real hidden issues in the new urban water 'political correctness' are the blockade of new supply and arbitrary rationing of existing supply."

8. Improperly maintained tanks also become havens for mosquitoes.

9. In early 2006 the Iemma government abandoned the previous nonconditional commitment to build a desalination plant, saying it would rely instead on taking extra water

from the Shoalhaven River, drawing on previously inaccessible water at the bottom of dams and announcing that reserves of groundwater in the Sydney region were larger than previously thought.

10. Rural-urban water trade could occur efficiently even if irrigators faced a monopoly urban buyer. That is because the water seller has options for selling it to other irrigators, using it himself, and, perhaps under future institutional arrangements, for selling it to environmental agencies with the power to trade in water.

11. One of a small number of recent developments in rural-urban water transfers is the announcement of a plan to pipe 75 gigalitres of water a year from the Goulburn Valley to Melbourne. This will require a pipeline estimated to cost $750M. It is planned that this water will represent one-third of water saved by investing $1Billion in upgrading irrigation infrastructure (Victorian Government 2007). The Productivity Commission (2006) and others have pointed out that infrastructure investments is often a high-cost way of obtaining extra water.

References

ABS (Australian Bureau of Statistics). 2004. *Water Account Australia 2000-0.* Cat.no. 4610.0, Canberra: ABS.

ABS (Australian Bureau of Statistics). 2005. 2003–2004 Household Expenditure Survey, ABS data available on request.

Baker, R. 2006. Water fight: $1.5bn plan sparks fury. *The Age*, September 13, p.1.

Banks, G. 2004. The good, the bad and the ugly: Economic perspectives on regulation in Australia. *Economic Papers* 23(1): 22–38.

Boland, J., and D. Whittington. 2000. The political economy of water tariff design in developing countries: Increasing block tariffs versus uniform price with rebate. In *The Political Economy of Water Reform,* edited by A. Dinar, 215–235. New York: Oxford University Press.

Business Council of Australia. 2006. *Water under pressure: Australia's man-made water scarcity and how to fix it,* BCA Report, September, Melbourne.

CoAG (Council of Australian Governments). 2004. Communique June 25, 2004. http://www.coag.gov.au/meetings/250604/ (accessed October 3, 2005).

Costello, P. 2005. *States urged to back national reform for continued growth.* Press Release No. 093, October 27.

Crase, L., and B. Dollery. 2006. Water rights: A comparison of the impacts of urban and irrigation reforms in Australia. *Australian Journal of Agricultural and Resource Economics* 50(3): 451–462.

Dwyer, G., P. Loke, S. Stone, and D. Peterson. 2005. Integrating rural and urban water markets in south east Australia: Preliminary analysis. Paper presented at OECD Workshop on Sustainability, Markets and Policies. November 14–18. Adelaide.

Dwyer, T. 2005. Conspiracy of silence over water charges. *Australian Financial Review,* January 6.

Dwyer, T. 2006. Urban water policy: In need of economics. *Agenda* 13(1): 3–16.

Edwards, G. 2005. Demand Management for Melbourne's Water. Paper presented at 34th Conference of Economists. September 26–28, University of Melbourne.

Edwards, G. 2006. Whose values count? Demand management for Melbourne's water. *Economic Record* 82 (Special Issue), 54–63.

ESC (Essential Services Commission). 2005. *Estimating Long Run Marginal Cost, Implications for Future Water Prices.* Information Paper. Melbourne: ESC, Melbourne.

Freebairn, J. 2003. Economic policy for rural and regional Australia. *Australian Journal of Agricultural and Resource Economics* 47(3): 389–414.

Freebairn, J. 2005. Early days with water markets. Paper presented at Industry Economic Conference. September 2005, La Trobe University.

Henstock, G. 2005. Observations on recent trends in water pricing. Paper presented at IIR Conference on Water Pricing. October 26–28, Melbourne.

Hodge, A., and M. Bachelard. 2005. Water transfers won't get PM's nod. *The Australian*, June 28, p.4.

Langford, J., and C. Piccinin. 2004. Institutional and regulatory arrangements in the Australian urban water industry. In Water and the Australian Economy. *Growth* 52:70–76.

Millis, N. 2004. Urban water cycle. In Water and the Australian economy. *Growth* 52:54–60.

National Water Commission. 2006. Progress on the National Water Initiative: A Report to the Council of Australian Governments, National Water Commission, June, Canberra.

Ng, Y.K. 1987. Equity, efficiency and financial viability: Public-utility pricing with special reference to water supply. *Australian Economic Review*. 3rd Quarter: 21–35.

Organization for Economic Cooperation and Development. 1999. *The price of water—Trends in OECD Countries*. Paris: OECD.

Pearce, D. 2005. Planning for future ACT water supplies—The role of cost-benefit analysis. Paper presented at Productivity Commission Seminar. November 9, Canberra.

Productivity Commission. 2005a. *Financial Performance of Government Trading Enterprises, 1999–00 to 2003–04,* Commission Research Paper, Canberra, July.

Productivity Commission. 2005b. *Review of National Competition Policy Reforms,* Report No. 33, Canberra.

Productivity Commission. 2006. *Rural Water Use and the Environment: The Role of Market Mechanisms.* Research Report, Melbourne, August.

Quiggin, J. 2006. Urban water supply in Australia: The option of diverting water from irrigation. *Public Policy* 1(1): 14–22.

Rogers, P., R. de Silva, and R. Bhatia. 2002. Water is an economic good: How to use prices to promote equity, efficiency and sustainability. *Water Policy* 4:1–17.

Sibly, H. 2006. Urban water pricing. *Agenda* 13:17–30.

Smith, B. 2006. 14,000 let the side down over water. *The Age*, September 2.

Sydney Water Corporation. 1999. *Annual Report 1998-99.*

Sydney Water Corporation (2005). http://www.sydneywater.com.au/OurSystemsAndOperations/WaterConsumptionStorageReport/ (accessed October 3, 2005).

Turnbull, M. 2006. Confronting our water challenge. Speech to the Brisbane Institute, July 25, Brisbane.

Victorian Government. 2004. *Securing our water future together.* White Paper, Victorian Government Department of Sustainability and Environment, Melbourne, June.

Victorian Government, 2007. *Our water our future: The next stage of the government's water plan.* Department of Sustainability and Environment, June.

WA Water. 2005. Water for Perth from the Kimberley. http://www.watercorporation.com.au/water_sources_kimberley.cfm (accessed November 30, 2005).

Wahlquist, A. 2005. Catch it if you can. *The Australian*, September 24.

WSAA (Water Services Association of Australia). 2005. *Testing the water, urban water in our growing cities: The risks, challenges, innovation and planning,* WSAA Position Paper No. 01, October.

WSAA and National Water Commission. 2007. *National performance report 2005–06, major urban water utilities.* WSAA, Melbourne

Watson, A. 2005. Competition and water: A curmudgeon's view. Paper presented at the Australian Competition and Consumer Commission conference Relationship between Essential Facilities and Downstream Markets. July 28, 2005, Gold Coast, Queensland.

Yarra Valley Water. 2004. *Annual Report.* Melbourne.

Young, M., W. Proctor, M. Ejaz Qureshi, and G. Wittwer. 2006. *Without water: The economics of supplying water to 5 million more Australians,* CSIRO, May.

CHAPTER 11

Acknowledging Scarcity and Achieving Reform

Lin Crase and Sue O'Keefe

*W*ATER POLICY IS A SOCIAL PHENOMENON insomuch as it represents a configuration of the established modes of human activity that govern the distribution of the resource. In addition, contemporary water policies and the institutional changes that have emerged in Australia over the past few decades are symptomatic of the realization that the costs attending the status quo exceed the likely costs of change. Treated in this way, several important questions emerge about the motivations for water reform and the processes adopted to accomplish it.

First, what was the information that made it unacceptable to maintain the current arrangements, and what factors prompted the necessity for reform? Second, given that water reform amounts to a reallocation of a scarce resource between competing uses and users, what techniques are available to accomplish this restructure? Third, if we accept that policy change is always costly, in one way or another, what are the types of costs that accompany each of the mechanisms for achieving the required redistribution? Fourth, what is the likely magnitude of these costs and how are they to be measured and compared, and by whom?

Answers to these fundamental questions provide a foundation for considering in greater detail the operation of water markets, the need for an adaptive approach on both policy and management fronts, and the role of technology in this context. Collectively, these topics provide an overview of the main coping strategies employed by governments to ameliorate the costs of policy change. In addition, this chapter provides useful groundwork for subsequent chapters in this book.

The remainder of this chapter is organized into four sections. In section 1 the function of information and its interaction with the reform processes are addressed. The information collection and dissemination processes employed in the *Living Murray* are explored and then contrasted with groundwater management planning in Western Australia. Section 2 offers a simple typology of reallocation mechanisms and positions these in the context of the objectives emanating from the various water planning processes in different states. The costs and benefits of each generic reform program are briefly examined in section 3 with examples drawn from the current initiatives. Finally, areas for potential policy improvement are identified in section 4, accompanied by some brief concluding remarks.

Information and the Motivation for Water Reform

Earlier in this book, a major transformation in Australian society, in the form of increased awareness of environmental degradation and a shift away from the developmentalist ethos that focused solely on the productive and extractive benefits of water resources, was noted. However, this fundamental shift is not evenly spread across all elements of the community. Moreover, the heterogeneity of preferences for enhanced environmental outcomes from water policy has fuelled acrimonious debate about the veracity of the scientific information that underpins decisionmaking (see, for instance, Marohasy 2003).

At a more general level, this reflects the practical difficulty of applying the notion of the precautionary principle in an environment where rights may be reassigned in an effort to provide a largely uncertain environmental improvement. The challenge confronting policymakers is to achieve the necessary redistribution of the resource in the face of criticism from those in the community with intense preferences for maintaining the status quo. Invariably, these groups resort to condemnation of the incomplete scientific information which points to the need for change, and offer their own competing interpretation of available data (see, for instance, Benson 2003).

In Australia, the National Strategy for Ecologically Sustainable Development (ESD) is the central document that guides policy formulation in these circumstances and incorporates most of the principles articulated in the Rio Declaration of 1992. McKay (Chapter 4) has already noted that there is considerable disquiet from a legal standpoint about defining ESD with precision. However, and despite the arguably noble ambitions that underpin this approach, two fundamental and important questions emerge from a water policy perspective: What is the state of scientific water knowledge in Australia? And how are the risk-weighted consequences of alternative policy responses to potential environmental harm to be assessed?

In relation to the first of these questions, elementary knowledge about the extent of national water resources has been found wanting. Answering questions like "how much water we have, where it is, where it is going, what it is being used for, and who is entitled to it" (NWC 2005, 1) requires baseline data that is still being assembled under the National Water Initiative. Clearly, our understanding of the present resource allocation is incomplete.

The significance of this knowledge gap should not be understated. By way of example, consider the level of understanding that confronts policymakers in

Western Australia, a state that relies heavily on groundwater for its water needs and is predicted to face amongst the most severe consequences of climate change (see, for instance, Pittock 2003, *1*). One of the major groundwater resources in this state is the Gnangara Mound located north of Perth. The Mound provides almost 60 percent of Perth's water needs, sustains a large horticultural region, and supports valuable ecosystems, such as cave systems and coastal lakes and wetlands. Notwithstanding the prominence of the Mound, and the marked recent decline in the availability of the resource, simply quantifying the rate of extraction has proven difficult. Until recently, only those users extracting 500 megaliters or more per year were required to report or meter their use. Clearly, "the lack of information on water use has hindered attempts to manage the problems of the Mound" (IRSC 2005, *15*).

In addition to the uncertainty circumscribing the quantum of water resources in numerous settings, there is a dearth of scientific knowledge about the relationship between water availability and the operation of ecosystems. Hillman (Chapter 9) provides evidence of attempts to remedy this situation. but much work remains to be done.

Notwithstanding these information deficiencies, the various levels of government have notionally applied the precautionary principle and commenced a program of reassigning water for environmental purposes, particularly where the evidence of environmental stress is convincing. To illustrate the variety of approaches adopted by policymakers, two specific cases are considered here: the *Living Murray* initiative and the assignment of environmental water provisions for the Gnangara Mound.

The Living Murray

In response to growing concerns about the environmental health of the River Murray, the Murray-Darling Basin Ministerial Council (MDBMC) released its *Living Murray* discussion paper in July 2002. The overriding purpose of the *Living Murray* process was to dedicate more water for environmental purposes and the document itself was designed to "start community discussion about whether or not water should be recovered from water users for the environment" (MDMC 2002, *29*). Three main reference points were proffered as a means of framing this discussion: namely, 350, 750, and 1,500 gigaliters.

The *Living Murray* followed other substantial reforms that had impacted the management of this iconic river, details of which are provided in earlier chapters in this volume. Significant policy events included the Cap on water diversions at 1993–1994 levels, after an audit in 1994 revealed significant growth in water extractions resulting in deleterious impacts on the riverine environment (DLWC 1997, *1*); the CoAG Agreement on Water Resource Policy (or Water Reform Framework) in February 1994, and later the Competition Principles Agreement in April 1995; and numerous legislative changes at the state level which were sympathetic to the thrust of the initial CoAG reforms.

The *Living Murray* was purportedly premised on community engagement processes under the auspices of the Independent Community Engagement Panel. As part of the *Living Murray* consultation process additional information was to be garnered by policymakers on several fronts. First, a scientific reference panel was assembled to provide information on the ecological consequences of alternative water allocation and flow regimes. The scientific reference panel set about investigating the ecological potential of the three reference points under three different operational scenarios. The first scenario represented the operational status quo (i.e., high summer flows for irrigation with extractions capped) whereas options b and c modeled the ecological benefits of flow regimes to target various ecosystem locations and ecological indicators (SRP—MDBC 2003, 6). Although the interim report provided "a basis for discussion amongst scientists, government officials and the broad community" (SRP—MDBC 2003, 6) the final report, which was scheduled for delivery in mid-2004, was not published by the MDBC.

The second genre of information to undergird the *Living Murray* related to the social impacts of reassigning additional water to achieve ecological changes. In this regard a social impact assessment framework was devised and tested but was not operationalized on a broader scale (see, for instance, EBC—MDBC 2003). Similarly, scoping of social impacts occurred but progress to stage 2 of a full social impact assessment never materialized.

The third form of information dealt specifically with the economic implications of assigning water to the environment. These analyses comprised two main types: investigation of the economic impacts on agriculture, given that most of the resource would need to be diverted from this sector, and the economic value of additional environmental flows. Poignantly, the latter study was to have comprised the application of a Choice Modeling technique to a wide cross-section of the community. This was to have included commentary on the ecological implications of the three reference points, some indication of the economic ramifications of resource reallocation and acknowledgment that the wider community would need to pay to achieve change. In effect, such a technique would have provided policymakers with a measure of the preferences of the wider community and simultaneously raised awareness of the consequences of any decision. This phase of the information gathering process was truncated by the bureaucratic and political players in the *Living Murray* debate. A prominent member of the MDBC at the time of these decisions later publicly mused that the bureaucracy was able to work out the community's preferences without the need for the information from a choice model (Chloe Munroe, February 12, 2004, pers. comm). Arguably, this also reflects the modest standing of techniques of this genre amongst the bureaucracy in Australia and contrasts with their wider acceptability in the United States and the United Kingdom. These actions also give some weight to Syme and Nancarrow's (Chapter 15) later claim that greater attention could be justifiably afforded to community preferences during the formulation of Australian water policy.

Importantly, the outcome from the *Living Murray*, referred to as *the first step*, became an integral component in the National Water Initiative with the announcement that member jurisdictions of the Murray-Darling Basin would allocate $500 million over the next five years to address the declining health of the rivers in the Basin, particularly the River Murray (CoAG 2003, 1).

Coincidentally, this represented the equivalent of about 500 gigaliters of water at current market rates, a quantum of water not considered as a reference point in any of the preceding scientific analysis.

These events provide a salient reminder that information gathering and water policy formulation do not operate in a linear manner with the former solely designed to enlighten the latter. Rather, vested interests and lobby groups may prefer that information remain imperfect for fear of undermining their influence over resource allocation. In addition, if the science is irrefutable and the preferences of the community well articulated, politicians and the bureaucracy must then run the risk of severe ridicule if they choose an allocation that aligns with their own ambitions rather than those of the broader community. Put simply, economically efficient outcomes that adequately encompass the full spectrum of community preferences should not be confused with those that are selected as being politically efficient at a particular point in the electoral cycle—particularly in the context of water.

The Gnangara Mound

As with the *Living Murray* decision, actions designed to redress the degradation of the Gnangara Mound have been circumscribed by information deficiencies. Thus, while "there is recognition of the need to take pressure off the Mound" there is also a lack of information capable of "[q]uantifying how much water needs to be saved and by whom." However, in line with the Intergovernmental Agreement on the Environment and the underlying ESD principles, the Western Australian water strategy argues that "the absence of this knowledge should not prevent action to save water from being taken now" (IRSC 2005, *15*). Thus, like the policymakers dealing with the decline of the River Murray, the West Australian experience is typified by progressive policy amendments as a reaction to emerging information.

It is possible to trace the evolution of the reassignment of Gnangara groundwater to the environment over several decades of information revelation and policy change. Groundwater extraction in this region began in earnest in the early 1970s. In 1987 an Environmental Review and Management Program was developed by the then Western Australian Water Authority, which placed some environmental conditions on further abstractions. As part of this policy, there was a necessity to develop a management and monitoring program which prompted increased ecological research into the impacts of altering the groundwater regime throughout the early 1990s. Findings from this research work were subsequently employed to determine the water requirements of wetlands that were dependent on the level of the groundwater in the Mound. In effect, the information from several studies into plants, aquatic invertebrates, and water birds was integrated to provide an indication of the requirements to prevent further degradation, manifested in maintenance of the current vegetation distributions (DEH 2001, *9*).

An important ingredient of this information–policy interaction was the establishment of Ecological Water Requirements (EWRs) and Environmental

Water Provisions (EWPs) as part of the statewide policy amendments of 2000. EWRs are defined as "the water regimes needed to maintain ecological values of water dependent ecosystems at a low level of risk" while EWPs represent "the water regimes that are provided as a result of the water allocation decision-making process" (WRC 2000, 2). Importantly, in the case of EWPs there is a necessity to take into account "ecological, social and economic impacts . . . [such that EWPs] may meet in part or in full the ecological water requirements" (WRC 2000, 2). Clearly, in this context there is a need for social and economic information to inform the formalization of EWPs.

In recognition of the absence of information to account for social and economic values and the potential conflict between the fulfillment of EWRs and the economic and social objectives, the Water and Rivers Commission placed considerable emphasis on community involvement as part of the determination process. In this instance community involvement in decisionmaking would have been conceptualized by the bureaucracy as being a less august process than that applied in the case of the *Living Murray*. In general, this amounted to the establishment of water resource management committees as a vehicle for engendering community/stakeholder involvement. Minimum requirements were established as part of statewide policy and encompassed "notification of the preparation of a draft plan; call for public submissions and the preparation of a summary document; referral of the plan, as modified as a result of submissions, to bodies which the Commission considers may be affected or should view the plan for any reason; and a further opportunity to provide submissions on the modified plan" (WRC 2000, *10*).

Clearly, the foreshadowed mechanisms for gathering additional information to inform policy decisions regarding the Gnangara Mound are less elaborate than those employed in the case of the *Living Murray*. At least two propositions account for this disparity. First, the proposed mechanisms in Western Australia may have been adjudged adequate to inform policymaking and avoid the potential loss of control by decisionmakers who might potentially be threatened by an overly informed electorate. A second explanation resides in the perceived importance of water scarcity at the time. More specifically, the Water and Rivers Commission observed that:

> While some eastern states are already in the situation where water has been over-allocated in many areas and there is an urgent need to greatly reduce consumptive use to provide more water for ecosystems, Western Australia is fortunate that this is not expected to be a major problem. (WRC 2000, *16*)

In reality, the optimistic outlook being offered by WRC at this time was itself ill-informed because of the significant funding constraints placed upon its activities. In 2003 the auditor general for Western Australia found that "WRC does not have the information they need to determine the sustainable use of the groundwater and surface water use in many areas of the state" (Auditor General of WA 2003, 3). It was noted that this deficiency was primarily due to the fact that state funding had "progressively declined from around $2 million per year in 1990 to around $300,000 in 2002" at a time of substantial increase in water demand.

This funding priority (or lack thereof) over this period stands in stark contrast to the more recent and increasing disquiet expressed in some quarters about the environmental sustainability of the Mound. For instance, Yesertener (2002) has noted a steady decline in the Superficial Aquifer storage between 1979 and 2000, while Vogwill (2004) shows this trend continuing between 2000 and 2003, thereby adversely affecting most wetlands (McCrea 2004). Reports that the level of groundwater storage has fallen by as much as twelve meters are common in the media (see, for instance, de Blas 2004) and a number of Water Corporation bores have been shut down in the areas of greatest concern (IRSC 2005, 5). Arguably, these events are as much a testament to government failure as the inadequacies of information at the time.

A Typology of Water Reallocation Mechanisms

The *Living Murray* and Gnangara Mound cases are illustrative of a broader trend in Australian water policy, which has seen increased emphasis on the necessity to redirect water resources to maintain or rehabilitate the health of water-related ecosystems. Similar examples can be found in most other states with the extent of extractive over-allocation varying in line with the degree of enthusiasm applied under the original developmentalist ethos described earlier by Musgrave (Chapter 3). Even in Tasmania, where diversions on average account for about 1 percent of the mean annual flow (NLWRA 2001), it has been necessary to put in place water use sustainability projects that attempt to deal with over-allocation problems in specific streams and aquifers (see, for example, DPIWE 2005, 46).[1] Guided by the themes expressed in the national agenda, most states and territories have now embarked on programs that have, as their stated goal, the redistribution of water to achieve environmental ends.

Confronted with the need to reallocate water for environmental or in-stream benefits, policymakers have two basic approaches to achieve this goal. First, all consumptive users could have their access rights unilaterally reduced to account for the needs of the environment. Second, water withdrawals undertaken by specific users can be reduced to achieve the necessary environmental gains. In the case of the latter approach, this might occur on the basis of some selective mandate by the state or, alternatively, it could be achieved through voluntary surrender of access rights using market instruments. As noted earlier in this book (see, for example, Crase and Dollery in Chapter 6), governments have been generally reluctant to reduce all access rights for fear that this would undermine the general sanctity of private rights. This has left policymakers choosing between state-selected projects for reducing water consumption and self-selected mechanisms via markets.

In many instances these programs have been euphemistically termed *water-recovery projects* or *water saving projects* (see, for example, Deamer 2005). In some cases the use of water markets to purchase water from willing sellers for environmental purposes has been included within this genre of strategy (see, for instance, IRSC 2005, *16*). However, the use of markets to recover water to achieve a redistribution for environmental ends has also attracted criticism from several quarters, and is frequently dismissed by some politicians and lobby groups as a last resort because of its potential to produce adverse impacts on industries and communities (see, for example, Miell 2003; Truss 2005). In line with this latter observation, it may therefore be more useful to categorize mechanisms designed to redistribute water to achieve environmental outcomes into two main forms: water recovery projects which operate independently of voluntary acquisition of water rights, and those activities with clear water market orientation. In addition, a range of solutions exist within the water recovery project typology and these broadly fall into engineering and managerial initiatives. It is also potentially feasible for some managerial initiatives to involve active participation in the water market. For example, options contracts might provide a vehicle for combining managerial and market mechanisms to achieve an environmental outcome (see, for example, Hafi et al. 2005).

State Water Recovery Projects

The defining characteristics of these projects are their reliance on achieving reallocation by nonmarket techniques and a strong involvement by the bureaucracy in decisionmaking. By and large, projects of this nature are aimed at "preventing losses between water leaving the dam (or aquifer) and use by plants or animals" (Deamer 2005, *1*). More generally, these projects concentrate on the technical efficiency of water use, particularly in irrigation.

The Murray-Darling Basin has provided the setting for numerous projects of this genre, primarily driven by the acute need for environmental reallocations in this region—diversions as a percentage of mean annual flow presently run at about 210 percent (NLWRA 2001). Deamer (2005, *4*) offers a review of water-saving projects in the Basin and groups them into eight main types: evaporation projects, metering, delivery system technology, piping supply on natural channels, piping supply on man-made channels, managing the flooding of wetlands, relocating excess water to the environment in wet years, and on-farm efficiencies. Notable examples of these types of initiatives include the piping of water for stock and domestic supplies in the Wimmera-Mallee in Victoria, and the decommissioning or reduced use of inefficient storages, like Menindee in New South Wales and Lake Mokoan in Victoria. Other projects include channel automation in irrigation districts in central Victoria, and the construction of levees to return ephemeral wetlands to their natural winter-spring flooding cycle. In essence, these works require two main ingredients which vary in relative importance between projects. First, there is an expanded engineering contribution to the mechanisms of water delivery and use. Second, projects of this type entail alternative management of water flows to meet environmental objectives. In sum, evaporation and seepage losses that formerly attended aged infrastructure or outdated management are saved and are then presumably garnered on behalf of the environment.

In order to achieve the coordination necessary to bring ventures of this nature to fruition, each of the state governments in the Basin has assigned bureaucracies the tasks of identifying and assessing potential projects. In addition, the New South Wales, Victorian, and Commonwealth governments collaboratively sponsored the formation of a private company, *Water for Rivers*, which has the goal of securing water to be returned as environmental flows to the Snowy River and River Murray. The combined environmental water target from the *Living Murray* initiative and the restoration of the Snowy River is 782 gigaliters over the next ten years.

Importantly, the financial cost of water recovery projects varies considerably, in both aggregate and per megaliter terms. For instance, for a modest public investment of $1 million in the Edward River Gulpa Island State Forest, simple engineering works have been employed to return the area to an ephemeral wetland. This investment simultaneously yielded over 19 gigaliters of environmental water at an average cost of only $52 per megaliter. By way of contrast, replacing the 17,500 kilometers of open earthen channel in the Wimmera-Mallee district of Victoria with pipes will cost over $500 million, save an estimated 103 gigaliters of water and yield environmental water at $4,860 per megaliter (Deamer 2005, 2–3).

On the basis of these data and the current market value of water rights held by irrigators, it would appear that financial costs represent only one criterion upon which competing water recovery projects are assessed by the bureaucracy. Permanent water access rights can presently be purchased in many catchments for far less than $4,000 per megaliter, often between $850 and $1,000 (ACIL Tasman 2003, 1). More specifically, the political acceptability of water recovery projects would appear to influence decisionmaking. The political appeal of water recovery projects stems from the perception that the wider costs of structural adjustments that attend such projects are less than those that attend the acquisition of access rights in a market setting.

There is also a property rights dimension to this issue which warrants discussion. Earlier, in Chapter 6, Crase and Dollery pointed out that although considerable effort had been put into defining the rights of some water users, elements of the water cycle remained ill-defined—return flows being a case in point. At a farm level increased irrigation efficiency tends to reduce return flows which, in many cases, underpin the base flow that provides environmental amenity. Engineering projects that purport to save water for the environment may well be a case of robbing Peter to pay Paul—except in this instance Paul might legally argue that Peter (return flows) was not technically robbed because the rights were ill-defined to begin with. Nevertheless, the political appeal of this approach has manifested in its widespread application in most jurisdictions.

Market Approaches

Notwithstanding the relative political strengths of state water recovery projects, market mechanisms remain an important element of the policy framework. Moreover, ACIL Tasman (2003, 1) observe in their report on the mechanisms for gaining environmental flows as part of the *Living Murray* initiative that "[t]he information available indicates that there are limited opportunities for water use efficiency savings at a marginal cost of less than $1,000/megaliter, except perhaps for reuse generally and for certain applications in horticulture." Similarly, and in the context of on-farm activities, ACIL Tasman (2003, *1*) contend that "it is also likely that those [projects] that are currently economic have been (or are being) implemented." Put simply, it seems unlikely that the quantum of water required to adequately restore environmental health will be achievable via the politically palatable state water recovery project options, at least in the case of the Murray-Darling Basin.

In the case of the Gnangara Mound discussed earlier in this chapter, recovery of water for environmental purposes is in its early stages. A recently announced government initiative allocated $29 million to enable Harvey Water (the main irrigation entity drawing water from the Mound) to replace open channels with a pipe network. Funding for these arrangements has been provided by the urban water authority (the Water Corporation) with water savings accruing to the urban Integrated Water Supply System. The Western Australian government has described these arrangements as a "trade agreement" (WAGPD 2005, *12*), despite lying within the state water recovery project genre in the context of the framework used here.

The willingness of the state to support engineering-centric water recovery projects, without resorting to market buy-back, is evidenced by the sponsorship of programs that attempt to artificially restore environmental water. For instance, the Department of Conservation and Land Management has recently undertaken a recovery project that involves refilling the Yanchep caves, located in the western precinct of the Mound. The caves have suffered from declining water levels since the 1970s and the project aims to ensure that the aquatic root mat that underpins the caves survives into the future. Paradoxically, the project involves the construction of an underground pipeline and pumping groundwater back into five of the caves in the area with the intention of creating "a localised ground water mound that will become self-sustaining" (CALM 2005, *1*). The willingness of governments to support projects of this nature in preference to using voluntary acquisition of water for environmental purposes is arguably indicative of the perceived political costs of activating market mechanisms in this context.

The Relative Merits of Alternative Policy Approaches

The penchant for water recovery projects and relative reluctance to engage in market activities to buy back water for environmental services has manifested itself in a funding distribution skewed toward the former approach within the National Water Initiative. In September 2004 the prime minister announced the establishment of a $2 billion Australian Government Water Fund. The fund comprises three main programs with one component, the Water Smart Australia Program, attracting most of the available funding ($1.6 billion). The aim of this program is to "accelerate the uptake of smart technologies and practices in water use across Australia . . . [with most support] directed to practical on-the-ground projects" (NWC 2005, *s1–1*). The Water Smart Australia Program is administered by the National Water Commission, which has listed nine main project types for which financial support is available. Notwithstanding that two of these project types have been assigned the broad headings of "improve river flows for better environmental outcomes" and "return groundwater aquifers to sustainable levels" (NWC 2005, *s2–1*), all remaining categories focus heavily on technical/managerial adjuncts for raising water use efficiency. For example, programs that lead to "improvements in irrigation infrastructure," or that "advance efficiency improvements on on-farm water use," or "develop water efficient housing design" (NWC 2005, *s2–1*) are all eligible for significant support. Tellingly, there would appear to be limited scope for market purchase of water from willing sellers under this program, regardless of the acknowledged limits to state water recovery projects (see, for instance, ACIL Tasman 2003).

In order to appreciate the enthusiasm for projects of this genre relative to the market alternative, it is necessary to review the merits and limitations of this policy approach. First, given that the majority of access rights to water resources are held by irrigators, any redistribution in favor of the environment requires this sector to forego or amend their consumptive behavior. State water recovery projects offer a low-transaction cost mechanism for coordinating behavioral adjustments within irrigation districts. Infrastructure and managerial changes of this magnitude are more easily coordinated at a superordinate level, thereby providing a policy justification for this approach.

Second, as has already been noted, projects of this type rarely attract high political costs. Those with strong claims on the resource (i.e., irrigators) are not required to relinquish those rights under water recovery projects.[2] Moreover, they often stand to gain an improved service, in the form of a more reliable piped infrastructure. In addition, while such projects may *prima facie* appear more costly than the market purchase of access rights, any additional costs are spread across a wider population who are seeking to improve the status of the environment. The strong claims of the few to shore up existing rights will usually mitigate against the new but modest claims of the many, in a political sense at least.

Third, in the absence of any metric to adjudge the value of environmental flows, policymakers can still effectively argue that the additional expenses for water recovery projects are justified by invoking a version of welfare-enhancing logic. Put simply, policymakers can spend $4,800 per megaliter (as opposed to $1,000) to garner environmental water because it is not possible to definitively prove that the community is unwilling to pay this amount, or more.

Fourth, water recovery projects avoid the complexity of third party effects that potentially attend market acquisition of access rights, or at least allow policymakers to ignore them. At least four types of third party impacts complicate the choice of a market regime: salinity impacts; the effect of return flows from irrigation users; the notion of stranded assets; and the impacts on system reliability (Brennan 2004, *18*). Water recovery projects avoid some of these externalities, although enhanced efficiencies in delivery infrastructure and on-farm use clearly have the potential to incidentally alter salinity and return flows (see Crase and Dollery, Chapter 6).

Notwithstanding the merits of water recovery projects, market-based techniques for acquiring environmental water also embody significant advantages. Brennan (2004, *1*) observed in her review of the economic issues in the *Living Murray* that "market mechanisms that inherently involve competition between water sellers that participate voluntarily will be a preferred approach to across-the-board reductions in water access rights, and this is justifiable both in terms of social acceptability (from the social impacts statement), and on economic efficiency grounds." Arguably, the economic and social justification of voluntary market mechanisms also translates into advantages over state water recovery projects.

Leaving aside the potential budgetary advantages of purchasing water at a lower cost than is facilitated by some water recovery projects, the rationale for supporting voluntary market exchange in this context can be traced to other factors.

First, a market setting for the purchase of water for environmental services would result in those who value access rights least, offering them for sale. Perhaps ironically, it was this same rationale (i.e. a water market moves the resource to its highest value use) that was used to underpin early CoAG reforms that broke the nexus between land and water rights, and the more recent National Water Initiative thrust for expanded trade and refinement of property rights. To date, however, trade to garner water for environmental purposes has largely been embraced only as a last resort and in some circumstances depicted as the taking of water from regional communities (see, for instance, Peatling 2003). Such a response clearly ignores the basics of mutually beneficial exchange that underpins a market framework.

Second, market acquisition of access rights provides scope for the government to take advantage of existing spatial constraints to water trade as a means of exercising price discrimination. In the case of the Murray-Darling Basin, spatial constraints on trade make it feasible to target the purchases of water from those areas with known low-value water uses. Brennan (2004, *17–18*) contends that "[p]erhaps most of the proposed benefits of a market approach to water acquisition will be able to be realized because of this spatial disparity in the value of irrigation water."

Third, there are significant informational advantages to employing market transactions to acquire water for environmental purposes. The market framework is conducive to an enhanced level of transparency on two fronts—it makes the benefits of exchange obvious to sellers and the budgetary costs of environmental decisions more apparent. The former information stands to improve irrigation communities' perceptions of the value of environmental services, insomuch as their members become the recipients of cash payments for access rights, some of which might be directed toward alternative local investments. In addition, the budgetary cost of environmental water acquisitions is clearer to the taxpayer. Unlike in the case of water recovery projects, employing extensive or elaborate infrastructure, it is difficult for bureaucrats to bundle a range of social objectives into a simple market acquisition of access rights.

Fourth, there are reasonable grounds to suggest that voluntary market acquisition is preferable on equity grounds. To reiterate, market acquisition is likely to be most appealing to those irrigators who value access rights least. Put differently, irrigation farmers who are unable to realize substantial profits from their current activities are more inclined to sell their access rights than are profitable farmers. By way of contrast, water recovery projects frequently enable profitable farmers to become more profitable and capture more of the resource through reuse and the like. In this context, localized rules that deliberately restrain trade become a vehicle for perpetuating existing income disparities within irrigation communities, while market acquisition of access rights offers a form of income redistribution.

Notwithstanding the merits of market-based transfer of access rights to achieve environmental ends, this approach remains contentious, in part, because of potential third-party effects and the inadequate definition of all rights to water. Important considerations in this context include the rights to return flows, water quality rights, the communal rights of shared irrigation infrastructure, and the impacts of trade and environmental diversions on the reliability of storages. The deficiencies with the rights to return flows and water quality have been addressed earlier by Crase and Dollery (Chapter 6). An attempt is made to briefly consider the third and fourth dimensions of trade externalities here.

McLeod and Warne (Chapter 7) provided a synoptic overview of the concerns widely held within the irrigation sector about the effects of permanent trade of water away from irrigation districts. These concerns revolve around the pecuniary externalities that would arise if some irrigators left the industry and the remaining farmers were then required to carry the costs of irrigation infrastructure across fewer farmers. The process of tagging accompanied by the implementation of exit fees have both been proffered as vehicles for dealing with these events. In effect, this represents community action to attenuate the rights of individual irrigators and might be regarded as an institutional manifestation of the stigmatization of water sellers (briefly described by Syme and Nancarrow in Chapter 15). Notwithstanding the social concern about water being exported from irrigation regions, pecuniary externalities provide only a weak economic case for attenuating property rights in this manner. It could be argued, for instance, that amending rights in this way is tantamount to profitable farmers exercising their right to enjoy continued subsidization at the expense of their poorer neighbors. Moreover, exit fees substantially mute the price signals that are required to indicate the true cost of resource use in a given context. Instead of imposing exit fees it may be more appropriate for irrigation companies to assist in sending signals that indicate those areas of an irrigation district which are most costly to maintain, rather than socializing the cost of irrigation across all users—both within and outside irrigation.[3] In any case, the economically appropriate mechanism for dealing with these issues is via other policies, such as taxation and social security measures, as noted by Edwards in Chapter 10.

A potentially more vexing issue relates to the impact of water trade (and environmental diversions) on the long term reliability of supply within catchments. Intervalley water trade has been permitted on a temporary basis since 1992 in New South Wales and, as noted by McLeod and Warne in Chapter 7, has resulted in significant annual transfers of water from the Murrumbidgee to the Murray catchments. Interstate temporary trade is also relatively commonplace. On November 1st, 2006, the New South Wales and South Australian governments took interstate trade to a new level by announcing that interstate permanent trade would be supported on the basis of tagged entitlements, whereby water traded out of a catchment remains subject to the rules and constraints imposed in the catchment of origin.

Tagging is an attempt to ensure that water trade does not unduly undermine the security of supply enjoyed in a given catchment. For instance, the hydrology of the Murrumbidgee Valley gives rise to a slightly higher reliability than that which attends the Murray. If water is freely traded from the Murrumbidee Valley to the Murray upstream of the confluence of the two streams this can potentially undermine the security of all users in the Murray upstream of Balranald (close to where the streams join). In effect, farmers in the Murray Valley may be making claims on water which is simply not in the supplying dam. Clearly, this is exacerbated by the activation of sleeper and dozer rights noted earlier by Quiggin in Chapter 5.

In recognition of this disparity, trading zones have been designed which prohibit, or limit, trades between areas of differing water supply reliabilities.[4] However, given the information deficiencies highlighted earlier in this chapter, this is an imprecise science. For instance, temporary trade from the Murrumbidgee to the Murray Irrigation Limited region (upstream of Balranald) faces no restriction. This occurs in spite of the known differences in the reliability of the two valleys. In years of reasonable rainfall the difference in reliability is considered insignificant and the transaction costs of making adjustments deemed too high. However, during exceedingly dry years this lack of precision can have significant impacts. On November 3rd, 2006, the Department of Natural Resources temporarily suspended all trade in the Murray and Murrumbidgee valleys. This announcement followed unprecedented low runoff into both catchments but stands as a poignant illustration of the potential nexus between water trade and supply reliability, when the rights to reliability are not comprehensively defined or are too costly to define.

A similar argument can be mounted in the case of water assigned to achieve environmental restoration. To date, much of the debate around environmental flows has focused on the quantum of water that might be ascribed to achieve environmental ends. However, as clearly illustrated by Hillman (Chapter 9) and Letcher and Powell (Chapter 2), an important ingredient of the restoration of natural flows is the timing of water releases from storage. Mimicking the natural flow of Australia's inland rivers ostensibly amounts to reducing the reliability of storages for other consumptive users and, if water is to be efficiently allocated to the environment via a market, the impacts on the reliability of supply must also be factored into the exchange. In Victoria, it has been foreshadowed that an environmental reserve be created and this is to be controlled by an environmental manager. At this stage this responsibility is likely to rest with regional catchment management authorities. As we have already noted, amongst the most efficient mechanisms for assembling this reserve is the purchase of water access rights from voluntary sellers. As it stands, however, neither party to this exchange has to account for how that water will be managed in the future. The disruption to conventional management of water infrastructure, which might accompany the transfer of the rights to an environmental manager, arguably undermines the rights of existing irrigators to an established level of supply security. Clearly, there is much work yet to be done before all dimensions of water property rights are defined and the market can weave its economic magic.

This creates a conundrum for water policymakers in Australia. On one hand, supporting and continuing constraints on the market stand to prolong the inefficiencies that are embedded within the status quo. Alternatively, embarking on a market free-for-all might be considered premature, and result in additional inefficiencies due to the imprecision with which rights are presently defined. Moreover, the costs of gaining additional information to create and enforce complete rights may be substantial. *Prima facie*, the policy response reflects this dilemma. The National Water Initiative espouses the virtues of the market and advocates stronger individual rights and a national water market; however, the efforts to recover water to achieve environmental amenity reflect a reluctance to progress wholeheartedly in this direction.

Current research effort is being directed at identifying alternative policy mechanisms that are capable of harnessing the benefits of market-based mechanisms without attracting third party effects. Work by ABARE into water trade that encapsulates salinity effects and the use of options contracts as a way of melding the benefits of trade with a solution to the stranded assets issue is illustrative of this approach (see, for example, Hafi et al. 2005). Unfortunately, a similar commitment to the market-based approaches is not yet evident at the political level, regardless of the caliber of the research work that points to its advantages.

Conclusion

The information that circumscribes water policy choices is incomplete in several ways. The deficiencies in our understanding of the ecology of Australian rivers and groundwater systems, and their relationship to extractive use, is matched by the paucity of information about society's preferences for funding environmental enhancements. By definition, this creates a complex choice environment where political leaders must recognize the limitations of science, and reluctantly acknowledge their own human constraints when interpreting emerging trends.

Notwithstanding the difficulty of making policy decisions in this climate, care needs to be taken to prevent rent seeking by interest groups, and others who might endeavor to exploit the information void to their own advantage. Arguably, instances of this type of behavior can already be found in recent water policy decisions.

One of the major challenges confronting policymakers is the desire to real-locate water resources to achieve environmental ends. While a range of strategies have emerged in this context, there is support for the view that many policymakers see the use of market-based approaches as a distinct genus of policy. Moreover, some political leaders have gone to extraordinary lengths to distance this approach from other water recovery projects: the latter being characterized by group or bureaucratic decisionmaking and a predilection for engineering and managerial mechanisms; the former being regarded as the policy of last resort.

The economic advantage of water recovery projects rests with the lower transaction costs of collective decisions and, in some instances, lower budget-ary costs to garner environmental water. In addition, these projects seldom attract political costs and, in this respect, represent the low-hanging fruit in the policy orchard.

By way of contrast, market-based mechanisms involving the voluntary acqui-sition of access rights have been received with less enthusiasm by policymakers to date. Regardless of the published advantages of this approach, and its docu-mented support in emerging legislation, there has been only limited deployment of this technique to acquire water for environmental services. Undoubtedly, this reflects the higher political costs attending this policy apparatus and the neces-sity to invest much greater effort in providing a more comprehensive definition of rights.

It remains to be seen whether the modest use of market-based approaches to gain water for environmental services persists as the marginal cost of water recovery projects rises to a point where it becomes discernable to the general public. Hopefully, the significant investment that has already been undertaken into encouraging research focused on innovative market-based mechanisms and the clarification of property rights will provide sufficient confidence for an expansion of this approach at a practical level in Australia.

Notes

1. The nuances created in Tasmania by the prominence of the hydroelectricity sector are detailed by Duncan and Kellow in Chapter 10.

2. Earlier, in McKay (Chapter 4), the process of establishing environmental claims on water was described. This was also given attention by Quiggin (Chapter 7). As a general rule, if water is set aside for environmental purposes as part of a management plan, compensation is not payable (although there are some variations on this approach across jurisdictions). Policy actions to deliberately recover water to meet an additional environmental target are treated as being over and above these claims.

3. It should be noted that differential pricing of water occurs in some irrigation districts to indicate the cost of delivery and maintenance varies. However, the flat rate exit fee being proposed in NSW irrigators makes no account of these differences.

4. Trading zones also endeavour to reflect the practical difficulties of delivering water in connected systems and, in some cases, the environmental consequences of increased water use in a given region (see Brennan 2006).

References

ACIL Tasman. 2003. Scope for Water Use Efficiency Savings as a Source of Water to meet increased Environmental Flows—Independent review. A Report to the Murray-Darling Basin Commission. Canberra: MDBC.

Auditor General for Western Australia. 2003. Second Public Sector Performance Report 2003. Perth: AGWA.

Benson, L. 2003. *The science behind the Living Murray*. Deniliquin: Murray Irrigation Ltd.

Brennan, D. 2004. Review of "Scoping of Economic Issues in the Living Murray, with an Emphasis on the Irrigations Sector." A report prepared for the Murray-Darling Basin Commission. Canberra: MDBC.

Brennan, D. 2006. Water policy reform in Australia: Lessons from the Victorian seasonal water market. *Australian Journal of Agricultural and Resource Economics* 50:403–423.

CALM (Department of Conservation and Land Management). 2005. Protecting Yanchep National Park Caves Fauna. News release, May 11, 2005.

CoAG (Council of Australian Governments). 2003. *Communique on the National Water Initiative*. Canberra: Council of Australian Governments.

Deamer, P. 2005. A review of water saving project costs and volumes. Paper presented at Australian National Council of Irrigation and Drainage (ANCID) Annual Conference. October 23–26, Mildura, Australia.

deBlas, A. 2004. Why the West is drying up. Earthbeat. Australia, ABC: Saturday 8.30. http://www.abc.net.au/rn/science/earth/stories/s1180408.htm (accessed November 1, 2005).

DEH (Department of Environment and Heritage). 2001. *Review of the Literature Describing Environmental Allocations for Wetlands. Environmental Water Requirements to Maintain Wetlands of National and International Importance*. Canberra: Department of Environment and Heritage.

DLWC (Department of Land and Water Conservation). 1997. Water audit—Implementation of the Diversion Cap in NSW. Paper presented to the Murray-Darling Basin Ministerial Council.

DPIWE (Department of Primary Industries Water and Environment—Tasmania). 2005. Report on the Operation of the Water Management Act 1999 (Tas.). Hobart: Department of Primary Industries Water and Environment.

Hafi, A., S. Beare, A. Heaney, and S. Page. 2005. Derivative options for environmental water. Paper presented at the 8th Annual AARES Symposium: Markets for Water—Prospects for WA. September 2005, Perth, Australia.

IRSC (Irrigation Review Steering Committee). 2005. *Irrigation review: Final report*. Perth: Government of Western Australia.

Marohasy, J. 2003. *Myths and the Murray: Measuring the Real State of the River Environment*. Melbourne: Institute of Public Affairs.

McCrea, A. 2004. Perth's groundwater mounds under pressure. Paper presented at Irrigation Association of Australia Biennial National Conference, May 2004, Adelaide, South Australia.

MDBC (Murray-Darling Basin Commission). 2002. *The Living Murray: A discussion paper on restoring the health of the River Murray*. Canberra: Murray-Darling Basin Ministerial Council.

Miell, D. 2003. *Truss on right track*. Media Release by NSW Irrigator's Council. Sydney, NSW Irrigator's Council. 2005.

NLWRA. 2001. *National land and water resources audit*. Canberra: National Heritage Trust.

NWC (National Water Commission). 2005. National Water Commission Annual Report 2004–2005. Canberra: National Water Commission.

Peatling, J. 2003. The federal government looks like side-stepping the hard decisions required to save the Murray River. *Sydney Morning Herald*, November 7. http://www.smh.com.au/articles/2003/11/07/1068013394570.html (accessed November 10, 2005).

Pittock, B. 2003. *Climate change—An Australian guide to the science and potential impacts.* Canberra: Department of Environment and Heritage.

SRP—MDBC (Scientific Reference Panel—Murray-Darling Basin Commission). 2003. Ecological Assessment of Environmental Flow Reference Points for the River Murray System: Interim Report. Canberra, Scientific Reference Panel—Murray-Darling Basin Commission.

Truss, W. 2005. *Coalition helps Snowy and River Murray flow.* News release, May 10, 2005.

Vogwill, R. 2004. *Groundwater modelling for the East Manneroo land and water use re-evaluation.* Hydrogeology report. Perth: Department of Environment.

WAGPD (West Australian Government Policy Divisions). 2005. *Response to the Irrigation Review Steering Committee.* Perth: West Australian Government Policy Divisions.

WRC (Water and Rivers Commission). 2000. *Environmental Water Provisions Policy for Western Australia.* Statewide policy No 5. Perth: Water and Rivers Commission.

Yesertener, C. 2002. Declining water levels on the Gnangara Mound. Water and Rivers Commission Report (unpublished).

CHAPTER 12

Urban Reuse and Desalination

Stuart Khan

THROUGHOUT MOST OF THE TWENTIETH CENTURY, engineered water management in Australia comprised dams to collect surface water, uncapped wells to extract groundwater, and outfalls to discharge primary or secondary treated effluents. As populations grew, more dams and wells were constructed and, in urban contexts, ever-increasing volumes of sewage were discharged into the country's waterways and surrounding seas. In this sense, water was treated as if it was a free resource with unlimited supplies. Little appreciation for the interdependence of all aspects of the water cycle was apparent.

However, as has already been observed in earlier chapters in this book, this approach is no longer consistent with the management ethos that underpins contemporary water policy—ideas about the role of technology are changing. More specifically, technology is being directed toward making more effective use of existing water, rather than harvesting additional water. This trend is particularly apparent in the urban water sector (see Edwards, Chapter 10). This chapter offers an overview of contemporary water technologies in the urban water sector and focuses specifically on water reuse and desalination. An effort is also made to provide a technological perspective of the present reforms and to highlight impediments to the expanded use of technologically based water solutions. Matters raised in this chapter should be considered against the contrasting economic backdrop provided by Edwards in Chapter 10.

The chapter itself is organized into five additional parts. The following section provides the context for recent urban water technologies and examines the process of preparing impaired water sources for urban use. Urban reuse and desalination are then considered separately in sections 2 and 3. The necessity for accompanying institutional changes is dealt with in the penultimate section before offering some brief concluding remarks.

The Context and Preparation of Impaired Water Sources

As a result of both population movements and recent droughts, most large cities in Australia entered the twenty-first century on the brink of requiring new urban water resources. For example, in 2002 a Commonwealth Parliament Senate Committee estimated that Perth would require new water sources by between 2005–2007; Brisbane by 2015; Canberra by around 2017, and Melbourne by 2040 (SECITARC 2002). Many of these predictions have since proved to be hopelessly optimistic.

The historical response to drought and population growth in Australia has been to increase storage capacity by building new dams on relatively untapped rivers. However, increased awareness of environmental and economic costs, coupled with the long-term inadequacy of many earmarked future dams, have caused governments to reconsider plans for their construction. The Victorian government has put in place policies aimed at deferring the need for a new dam for Melbourne for at least 50 years (Victoria DSE 2004). The New South Wales (NSW) government has announced an indefinite deferral of a long-planned additional dam for Sydney (NSW DIPNR 2004). It is likely that this will be the future trend for many large dam proposals around Australia as their environmental implications are increasingly recognized.

Adequate clean water is not the only challenge confronting urban communities. Marine pollution caused by sewage discharge is recognized as a growing worldwide problem (GESAMP 1998). According to the Clean Oceans Foundation, Australia has 142 estuary and ocean outfalls discharging more than a billion megaliters of effluent every year. Some of the largest outfalls discharge after only primary treatment.

The range of toxic chemicals known to persist in primary and secondary treated sewages includes heavy metals, aromatic hydrocarbons, and organochlorine compounds. For example, 39 tons of chlorophenols were reported to be discharged off the coast of Brisbane during 2004–2005 (Australian DEH 2006). Estrogenic steroid hormones (including estrone, estradiol, and the synthetic ethynylestradiol) have been reported in sediments adjacent to one of Sydney's largest ocean outfalls (Feitz et al. 2005). This suggests that these compounds, associated with particulates in the sewage, aggregate on contact with high-ionic-strength seawater and accumulate on the sea floor. Public health may also be at risk from the environmental presence of pathogenic organisms such as bacteria and viruses.

The Federal Parliament's Senate Environment Communications Information Technology and the Arts References Committee (SECITARC 2002: *16–17*) was established to inquire into Australia's management of urban water and reported that there were major opportunities for Australia to improve on its performance with regard to water reuse. It observed that in Australian cities "efficient water use is still perceived as an emergency measure to be adopted during drought condition." It asserted that "in a country of such limited water resources, this behavior must be the norm, not the exception." Among the recommendations of the Senate Committee was that "Australians generally be encouraged and assisted to use less water, recycle more effluent and significantly reduce the impact that urban development and its storm water collection and transport has on natural systems."

In 2003 the prime minister's Science, Engineering and Innovation Council (PMSEIC) identified possible mechanisms by which Australian cities could make better use of available water resources (Rathjen et al. 2003). The PMSEIC indicated that a mixture of initiatives appropriate to specific circumstances of each city would be required. However, essential criteria for all initiatives would include maintenance of public health, economic viability, environmental sustainability, and social acceptance. With these criteria in mind, recycled water was promoted by the PMSEIC as a valuable resource that should not be wasted and which can be used in a safe and sustainable manner to reduce pressures on limited drinking water resources. Thus, political interest in recycling and reusing water predates the 2004 intergovernmental agreement on a National Water Initiative (NWI) which, amongst others, emphasizes the development of innovation and capacity building to create water sensitive Australian cities as a matter of urgency.

Preparing Impaired Water Sources for Use

Although it is not appropriate to provide a detailed description of water treatment technologies here, a broad overview is presented as a foundation for subsequent analyses of the issues associated with the implementation of existing and emerging technologies for water management.

Advanced biological treatment processes most commonly rely on the expanded employment of microorganisms for the degradation and/or assimilation of chemical contaminants. Most notably, the use of anaerobic and anoxic conditions for processes such as denitrification have greatly expanded the range of treatable contaminants compared to traditional aerobic processes. An approach particularly suited to many advanced water treatment schemes is known as biological activated carbon filtration. This process involves the percolation of water through a granular activated carbon system on which a heavy biofilm has been established. Although since the activated carbon retains contaminants by adsorption, organisms in the biofilm are provided with increased opportunity to degrade them.

Chemical disinfection and treatment of organics is most commonly undertaken with oxidants such as chlorine or ozone. These processes may result in a direct molecular degradation of the target molecules, or render by-products more amenable to a secondary physical or biological removal step. Chemical treatment processes can be highly effective, however in some cases they can also be expensive to install and operate. Because they degrade, rather than remove contaminants, further issues arise with degradation products and by-products which may, in some instances, be of greater concern than the initial contaminants.

Photochemical disinfection and degradation of organics may be induced by exposure to natural sunlight or facilitated by an ultraviolet (UV) radiation source. The primary mechanism of UV disinfection is to initiate photochemical reactions, which effectively damage DNA molecules so that cell reproduction can no longer occur. Photochemical treatment relies on low turbidity in the water, to which recycled waters do not always conform. UV degradation of organics is still an emerging technology, however very high dosages are likely to be required for the removal of some recalcitrant species.

Physical methods of removing chemical and microbial components have traditionally relied on adsorption of target contaminants on either fixed solid surfaces (as in sand or granular activated carbon filtration) or on suspended particulates such as iron or aluminum oxyhydroxides or powdered activated carbon. Some further advanced physical treatment processes rely more on size-exclusion and electrostatic repulsion than simply on adsorption processes, and hence present a more reliable barrier. Membrane filtration processes (e.g., microfiltration, nanofiltration, and reverse osmosis) may result in significantly improved treatment of some key compounds. These processes, however, may be prone to fouling and can be highly energy intensive for some applications.

Much public discussion has taken place in recent years regarding the relative energy requirements to treat municipal wastewater and seawater to qualities suitable for reuse. To achieve best-quality water production, a number of alternative treatment approaches could be considered. However, currently in Australia, reverse osmosis membrane treatment is by far the most energy efficient approach for adequately upgrading both conventionally treated wastewater and seawater. Reverse osmosis technology has developed dramatically during the last decade, decreasing the energy costs and therefore the financial costs of treatment. However, the major source of energy requirement remains the necessity to overcome the osmotic potential difference across the membrane, that is, the difference in salinity between the purified water and the retained brine. Seawater is roughly an order of magnitude more saline than most conventional wastewaters. Accordingly, as long as the need to overcome osmotic potential remains the principal source of energy requirement, reverse osmosis of seawater will remain more energy intensive than the same treatment applied to conventionally treated wastewater.

Municipal Water Reuse: Current Practices and Future Prospects

In 2004, the Australian Academy of Technological Sciences and Engineering (AATSE) published a comprehensive report detailing current practices of water recycling in Australia (Radcliffe 2004). This report identified substantial variations in the relative proportions of available municipal sewage effluents that were reused by the various states and territories in 2001–2002. Significantly, much greater rates of water reuse were reported for each overall state than for the respective state capitals. This reflects the greater rates of water recycling in rural towns, particularly those in inland areas.

The AATSE report indicated considerable diversity in the approaches taken to reuse water throughout Australia which arguably provides a considerable knowledge and experience base which will benefit planning for future schemes. The principal approaches to water reuse in Australia are summarized here.

Onsite Municipal Reuse

Onsite municipal water reuse is practiced in Australia primarily by the selective capture of grey water sources from laundries and bathrooms. Typically, the grey water is treated by sand filtration and reused for toilet flushing and garden watering.

Very few houses and offices in Australia are capable of treating and reusing black water sources (such as from toilet flushings). Such systems require biological amelioration and disinfection. The few systems in existence operate primarily as experimental or demonstration schemes since they are expensive to install and require careful ongoing management.

Targeted Municipal Irrigation

Targeted municipal irrigation schemes are among the most common means of water reuse in Australia. In many cases, secondary or tertiary treated effluent is applied to public parks and gardens, golf courses, and playing fields. Such reuse practices are attractive primarily for the generally low levels of treatment required and the need for a relatively small number of distribution pipes to transport the water to the points of use.

An alternative approach has been developed with the introduction of small portable sewer-mining operations. These involve the extraction of untreated sewage from municipal sewer mains. The water is then treated by a small, sometimes mobile, treatment plant (normally using membrane technology) and reused for irrigation. An advantage of portable sewer mining operations is that they may be relocated depending on temporary or seasonal demands. Sewer mines have been trialed in various public locations around Melbourne, Sydney, and Canberra.

Industrial Reuse

Brisbane Water operates a successful industrial reuse program from the Luggage Point Sewage Treatment Plant. This plant delivers ten to fifteen megaliters of treated effluent per day to the adjacent BP Amoco oil refinery where is it used as boiler feed water. In addition to the potable water savings, this scheme allows for considerable infrastructure savings by eliminating the need to expand potable water mains capacity. There are now plans to increase the use of recycled water from Luggage Point. A major component of the Western Corridor Recycled Water Project is to supply water to Tarong and Swanbank power stations, which are major suppliers of electricity to Brisbane.

A large industrial scheme has been initiated in Wollongong (NSW), where Sydney Water now provides Bluescope Steel with twenty megaliters of high quality municipal recycled water per day for use in the steel manufacturing plant, comprising more than half of the plant's total water requirements. Further large-scale agreements for industrial reuse operations have been implemented at Kwinana (WA) involving mining, power generation, chemical fertilizer, and petroleum companies.

Agricultural Reuse

Recycled water from Adelaide's largest water treatment plant is delivered via the Virginia Pipeline to agricultural areas on the Northern Adelaide Plains and the Barossa Valley. The scheme supports one of Australia's most valuable produce markets and provides an alternative source of water to the over-utilized local groundwater. The Virginia Pipeline scheme was commissioned in 1999 and has a capacity of 110 megaliters per day delivered via a network of more than 100 kilometers of pipes.

A proposal for a pipeline from Brisbane, to supply recycled water to the Lockyer Valley and Darling Downs, was once deemed economically and environmentally unviable. However, due to ongoing water shortages, some such irrigation capacity may be provided by the Western Corridor Recycled Water Project, currently under construction by the Queensland government. A similar scheme has been initiated to deliver water from Hobart to the Coal River Valley in Tasmania. Horticulture and pasture irrigation are the focus of major plans to expand water reuse in Victoria over the next couple of decades. Agricultural reuse has also been successfully practiced by a number of much smaller applications, such as the Gerringong-Gerroa sewerage scheme, and Shoalhaven Water's Reclaimed Water Management Scheme.

Reticulation for Household Reuse

A small but growing number of new housing development areas in Australia have incorporated dual reticulation systems for the redistribution of treated sewage back to households. These comprise a dedicated system of pipes, taps and fittings, which must be kept entirely segregated from the potable water supply and out-going sewage mains. The water delivered by dual reticulation schemes may only be used for a limited range of applications such as toilet flushing and garden watering.

The largest dual-reticulation scheme in Australia began operation at Rouse Hill (NSW) in 2001. The scheme is continuing to expand and currently services more than 25,000 properties. The over-riding purpose of the Rouse Hill scheme was to protect the Hawkesbury-Nepean river system from the environmental impact of increasing urban development. More recently, dual-reticulation schemes have been established at Newington (NSW), Mawson Lakes (SA), Epping North (VIC), and Springfield (QLD).

Stream Flow Augmentation

Stream flow augmentation is among the most common but least-recognizable forms of water reuse. Effluent discharge into waterways is widespread in Australia, but does not always impart environmental benefits in terms of flow augmentation; however, there are cases where suitably treated discharges may have positive environmental impacts when released to waterways in carefully controlled flow regimes. If managed appropriately, stream flow augmentation can be environmentally beneficial but, with the some important exceptions, it rarely relieves pressure on potable water supplies. Planning is currently underway in Western Sydney to upgrade treated effluent from three large sewage treatment plants (at Penrith, St. Marys, and Quakers Hill). The reverse-osmosis-treated water is to be used to supplement flows in the Nepean River (at Penrith). This scheme will allow considerable savings of drinking water supplies by replacing those which are currently released from Warragamba Dam to maintain adequate river flow.

Indirect Potable Reuse

Contrary to many other parts of the world, unplanned indirect potable reuse is not widely practiced on a large scale in Australia. This is, in part, due to Australia's population distribution where most of the largest cities are in coastal areas at the bottom of their relevant catchment systems.

Nonetheless, there has been a rapidly increasing interest in planned indirect potable recycling (IPR) schemes since 2005. The Commonwealth government's National Water Commission (see McKay, Chapter 4) has provided the opportunity for some cities to investigate and cofund potential planned IPR schemes. A proposed IPR scheme was rejected by a community poll in Toowoomba (QLD) during 2006. However, a very large IPR scheme, known as the Western Corridor Recycled Water Project, is currently under development in South East Queensland. This scheme involves the treated effluents of six sewage treatment plants, which will be further treated at three new advanced water treatment plants. Some of the water produced will be used to replenish supplies in South East Queensland's largest reservoir, Lake Wivenhoe.

The Western Australian government is currently assessing the feasibility of recharging aquifers for indirect supplementation of Perth's water supplies. This approach, which involves an underground residence time of about a decade, may prove to be an effective means of overcoming psychological barriers to the concept of indirect potable recycling.

Future Prospects for Water Reuse

In 2004, the Council of Australian Governments (CoAG) included measures to encourage cost effective reuse and recycling of wastewater to urban water reform in the National Water Initiative (CoAG 2004). Most state and territory governments have now announced targets or estimated potentials to increase rates of water reuse in their capital cites. These targets and estimated potentials are summarized in Table 12–1.

Although households account for only around nine percent of water consumption, there remains significant potential demand for recycled water in urban areas. Around eight percent of household water is used in kitchens for consumption and food preparation. A further 20 percent is used in bathrooms, also involving intimate

Table 12-1. *Future Water Recycling Targets of Capital Cities*

State/Territory Capital	Future recycling targets or estimated potentials	Reference
Sydney	4% by 2010 (22 billion litres, target)	(NSW DIPNR 2004)
Melbourne	20% by 2010 (target)	(Victoria DSE 2004)
Canberra	20% by 2013 (target)	(ACT Environment ACT 2004)
Adelaide	33% by 2025 (estimated potential)	(SA Government 2004)
Perth	20% by 2012 (target)	(WA Government 2004)

contact. However, the remainder presents an opportunity for much of the potable water usage to be substituted with reuse even where potable recycling schemes are not preferred. A major advantage of household reuse is that the demand location is typically very close to the supply location. This applies to water sourced from municipal treatment plants and even more so for on-site reuse systems.

Agricultural industries are by far the largest consumers of water in Australia. In theory, every drop generated from municipal sewage could be reused for a secondary agricultural use, thus providing barely 10 percent of the water consumed for agricultural applications. The major difficulty is that large agricultural schemes are often located considerable distances from large sources of recycled water.

Mining ventures can require very large volumes of water, often in very concentrated areas. A number of large cities, such as Newcastle and Wollongong in NSW have extensive coal mining activities located close to sources of municipal wastewaters. Other mines are located in very isolated, but very dry, regions such as Kalgoorlie and Boulder in Western Australia.

Many manufacturing industries are high value industries for which reliable supplies of large volumes of very clean water are a crucial resource. For example, the need for a sure supply of highly purified water for electronics manufacturing was a major driver for water reuse in Singapore. High value manufacturing applications provide a strong incentive to apply effective treatment processes that may raise the cost of water treatment. Because water needs vary considerably from industry to industry, and from application to application, identifying suitable opportunities for reuse will not be a trivial task.

Gas and electricity industries may require a lower grade of water compared with many manufacturing industries and may also be more concentrated in their points of water use. The successful use of recycled water at Pacific Power's Eraring Power Station in NSW provides an example of the mutually beneficial results that can be achieved when water and power sectors cooperate. This is a precedent that could be repeated many times over in Australia.

Sewerage industries are perhaps the most opportune environments for water reuse practices. Unlimited quantities of treated wastewater make reuse an ideal option for many sewage treatment plant applications, such as cleaning operations. Quality considerations are unlikely to be a significant factor for many sewage treatment plant reuse applications.

In promoting water reuse practices, the identification of potential uses of recycled water is merely a first step. Before large-scale reuse can eventuate in all cities, changes may be required in the management and regulation of water resources. For example, the advantages of streamlined chains of responsibility for potable water, sewage, and storm water have been promoted (Hatton, Mac-Donald, and Dyack 2004). Such integrated water management would naturally lead to more transparent costing and pricing structures for water. For example, it would facilitate the objective of including costs associated with current externalities such as sewage treatment into pricing schemes for potable water delivery. This would enable a more realistic comparison of the costs of recycled water and fresh water supplies.

The Emerging Role for Desalination

Prior to 2004, desalination was practiced in Australia with just a few very small brackish groundwater schemes. The Premier of NSW had disparagingly referred to desalinated seawater as bottled electricity, noting the considerable energy requirements for its production. However, serious consideration of large-scale seawater desalination schemes began to accelerate during 2004–2006. This change of attitude was initiated by the severe worsening of some large-city water supply shortages, concurrent with developments in reverse osmosis technology, resulting in improved energy efficiency and considerably reduced treatment costs.

In December 2006, the Western Australian government announced the opening of Australia's first seawater desalination plant to supply Perth with up to 45 gigaliters per year of potable water. Soon after, the NSW government began construction of a desalination plant for Sydney on the Kurnell peninsular. Other cities, including the Gold Coast and numerous smaller coastal towns around Australia have also begun planning for seawater desalination as a component of their overall municipal water supply and management.

The increased national interest in desalination has been reflected by significantly increased scrutiny of the less desirable consequences. One such consequence is widely considered to be a weakening of the message highlighting the importance of water conservation. When a potential water source is envisaged to be as great as the world's oceans, the argument goes, the urgency to implement water-efficient technologies and practices is reduced. According to Sydney Water, during the period of mandatory water restrictions October 2003 to December 2005, total water consumption was 12.6 percent below the ten-year average (Sydney Water 2005). This is greater than the volume deliverable by the planned desalination plant (9 percent of the cities daily needs).[1] Furthermore, cities that come to rely on seawater desalination, rather than conservation or recycling, will also rely on ocean outfall infrastructure for the discharge of municipal wastewaters. As these cities harvest ever-increasing volumes of water from the ocean, they must also discharge similarly increasing volumes to the detriment of the marine environment.

The energy requirements associated with seawater desalination derive from both the treatment processes (typically reverse osmosis) and the need to pump the water uphill from sea level to a height sufficient for gravity-fed reticulation. In almost all Australian cases, these large energy costs will be met by coal-combustion electricity production. However, some of the larger schemes have sought to offset the associated greenhouse gas implications by effective carbon trading. For example, while constructing the Perth desalination plant, the Western Australian government concurrently invested in the use of windfarms to produce an equivalent quantity of energy to be fed back into the state electricity grid.

The fundamental principal of reverse osmosis is the employment of semipermeable membranes to separate a purified component of the water from a wastestream retaining the concentrated salts. This waste stream is commonly referred to as the membrane concentrate or brine. The sound management and disposal of concentrates has become one of the greatest concerns regarding water reuse and desalination, and is often a key factor determining the overall viability of a project. The issues involved include technical challenges, permitting problems, and high costs. Concentrate from seawater desalination typically comprises half of the original in-take volume and practically all of the dissolved salts. Accordingly, it is typically double the normal concentration of seawater. Most commonly, concentrates are discharged via ocean outfalls, however the double salinity renders concentrate plumes denser than seawater and thus they can be difficult to disperse. The potential impact of concentrate plumes on marine species in Australian environments has yet to be properly assessed.

One approach to seawater desalination that has quickly grown in popularity in the United States, and may be applicable in Australia, is that of colocation of desalination plants with power generation plants. Colocation has a number of advantages with the potential to make some progress toward overcoming many of the difficulties associated with seawater desalination. The key feature of colocation is the direct connection of the desalination plant intake and discharge facilities to the discharge outfall of an adjacently located coastal power generation plant. This provides for the use of the power plant cooling water both as a source water for the seawater desalination plant and as a blending water to reduce the salinity of the desalination concentrate prior to the discharge to the ocean. The advantages include reduced impact on marine environment as a result of faster dissipation of thermal plume and (diluted) concentrate.

Improvements and Reforms Required for an Increased Role for Technology-Based Solutions

The identified drivers for increased application of technology-based approaches to water management indicate an urgent need for such practices, particularly water recycling, to be more widely considered within Australia. Although technologically advanced initiatives are indeed increasing, a number of institutional reforms and improved practices will be required to achieve a significant degree of growth. The major areas requiring urgent attention in Australia are discussed here.

Financial and Economic Reform to Facilitate Technological Implementation

A narrow financial analysis of many Australian reuse and desalination proposals may often appear highly unfavorable. Typically, this is the result of the high infrastructure and operational costs associated with most advanced water treatment and distribution systems. Furthermore, revenue potential may appear low because nonpotable recycled water is typically priced at a lower rate than that which consumers are charged for traditional potable supplies. Revenue projections in some areas may also suffer from a limited market of potential applications and customers for nonpotable water supplies. Accordingly, it is often difficult for agencies to make a business case for reuse and desalination projects based solely on an assessment of the internal financial outcomes.

Financial assessments are, of course, very important and are useful for many applications. However, it has been argued that their scope provides too limited a context within which to evaluate the real social worth of many alternative water management options (Raucher 2005). This is because such assessments focus strictly on revenue and cost streams internal to the water agency, and these internal cash flows are not the same as the true value of most schemes to the greater community and society as a whole. For example, a financial assessment does not include the social benefits of sustaining an agricultural community during times of limited alternative water supply availability.

Unfortunately, many of the benefits arising from a reuse or desalination project can be difficult to identify and quantify. One reason is that the benefits are often very diverse in type and some may not be immediately obvious to some parties. Many of the benefits can be difficult to explain or to estimate in monetary terms. Furthermore, there is potential decoupling of beneficiaries from those who pay because water agency and political boundaries may not always be well aligned. To address these difficulties, it is necessary to conduct some form of full social cost accounting.

Recycled water costs may be distributed between water and wastewater users, both through capital facility charges and usage rates. Carefully applied cost allocations can allow for economic incentives to attract current or future potable water users to replace some of this use with nonpotable recycled water. The greatest challenge is to identify the most effective balance between pricing recycled water such that it is attractive and maintaining a nexus with the cost of service.

Tertiary treatment costs have historically been recovered from wastewater users (or generators), spreading the costs over the treatment authority's entire customer base. This approach has been logical because much of this tertiary treatment has been to facilitate a means of effluent disposal and thus represents a cost for managing wastewater. In many future situations, however, tertiary treatment may be implemented primarily to provide recycled water to offset demand on the drinking water supply. In such cases, the cost of tertiary treatment might logically be partially recovered from drinking water customers who may not be the same as the wastewater customers and may or may not be serviced by the same organization.

A major obstacle to water reuse in Australia has been the widely acknowledged historic under-valuing and under-pricing of fresh water supplies (Rathjen et al. 2003; Hatton MacDonald and Dyack 2004; Khan et al. 2004; Radcliffe 2004; WSAA 2005). The relatively small financial costs incurred in the use and disposal of fresh water supplies have provided little market-force incentive to embrace the less competitive water recycling applications.

Municipal potable water generally costs Australian consumers slightly more than $1 per kilolitre (a description of the rates applied in different Australian cities is offered by Edwards in Chapter 10). The low cost is partly the result of there being no requirement in pricing regimes to include costs associated with catchment management and protection of effluent-receiving environments (SECITARC 2002). In many cases, the consumer also pays sewerage charges that include the cost of treatment to a standard that is acceptable for discharge to the environment. These are separately accounted for and not integrated with the costs associated with producing and delivering potable water. The cost of producing and delivering recycled water is generally greater than the costs for fresh water; however, users are typically charged less for recycled water than for fresh water due to its more limited use.

As part of the National Water Initiative, Australian governments have agreed in principle to the implementation of water pricing and institutional arrangements which promote economically efficient and sustainable use of water resources (CoAG 2004). As noted by Edwards (Chapter 10), this has resulted in the use of "inclining block tariffs" in most jurisdictions. Alternative water-pricing frameworks have been proposed by both the Australian Water Conservation and Reuse Research Program (Hatton MacDonald 2004) and the Water Services Association of Australia (WSAA 2005). These whole water cycle approaches acknowledge the importance of applying consistent consideration of externalities associated with potable water, reuse water, and sewage. Both proposed frameworks include the provision of price signals that reflect the scarcity of resources as well as the costs associated with water treatment and delivery.

Advanced Understanding of Community Acceptance

Community concerns will significantly impact on the way water is ultimately managed in Australia, as they have in other countries (Po et al. 2004). A lack of community acceptance prevented an early indirect potable reuse proposal from proceeding in Caboolture (QLD) during the 1990s (NRMMC 2005). The political fallout attributed to this incident has bred some reluctance among other governing bodies to make policy statements on water recycling without assurances of community support. The political ramifications of a similarly unpopular proposal in Toowoomba (QLD) during 2006 are yet to be fully determined, but are unlikely to instill renewed confidence.

A survey of residents in a development area incorporating a dual reticulation system indicated strong support for some household water reuse applications, such as watering lawns and gardens. However, the support dropped off sharply as proposed uses of the water became more personal. Only moderate support was reported for clothes washing and extremely low support for the supplementation of drinking water (McKay and Hurlimann 2003). These attitudes will play a very strong role in determining the specific applications for which recycled water will be used and those that are currently unacceptable to some communities.

In addition to end-use applications, community attitudes will impact on the triple bottom line viability of water reuse schemes. For example, a qualitative study of Australian households connected to dual reticulation systems found that few would be willing to pay more for water as a conservation measure. The study participants also generally agreed that non-potable recycled water should cost less than potable water due to its more limited uses (Marks et al. 2003).

It may be expected that desalination schemes will prove to be a significantly easier sell to the community than many water reuse schemes, particularly potable water reuse schemes. However, the experiences of the NSW state government during 2005 and 2006 have demonstrated that considerable community opposition to major desalination schemes can also be generated or manifest in some circumstances.

Advanced Understanding of Industry Acceptance

There is evidence that progress must also be made toward improved understandings with some industry groups, rightfully concerned for the safety of their members and security of their industries.

For example, in May 2005 the Queensland branch of the United Firefighters Union of Australia released a statement rejecting the use of recycled water for firefighting purposes "unless there can be assurances that its use will cause no long term or short term ill effects to firefighters" (United Firefighters Union of Australia 2005). Thorough reviews and risk assessments have concluded that properly treated and managed recycled water would be acceptably safe for firefighting (WSAA 2004). However, without formal assurances, such union bans have the potential to significantly impact many water recycling schemes.

The Australian horticultural industry has also expressed trepidation toward the use of recycled water for irrigation (Hamilton et al. 2005). The major concerns are reported to be insufficient knowledge of impacts on market access; commitment to provide continuity of quality and supply to markets; implications of substitution of alternative water sources on security of supply; insufficient knowledge of food safety issues; inadequate understanding of consumer perceptions; and uncertainty about pricing of reclaimed water.

Improved Definition of Infrastructure Access Rights for Private Enterprise

Failure to recognize the potential value of recyclable sewage streams and to define access rights to them has been an impediment to some potential water reuse activities (Hatton MacDonald and Dyack 2004). During the 1990s, the Commonwealth government established means for private organizations to seek access to publicly owned infrastructure under reasonable terms and conditions. The Independent Pricing and Regulatory Tribunal of New South Wales has also recommended the introduction of competitive reform to Sydney's water industry (IPART 2005). However, little use has been made of these provisions to date and the few significant applications have met with initial resistance from state governments.

Since 1999 a private organization has petitioned the NSW government with a proposal to provide an alternative sewage service in Sydney. This would involve treating a large proportion of Sydney's sewage and then piping it west, where it would be used in agriculture and to supplement environmental river flows. The firm would make its money by offering customers an alternative sewage service, selling the reclaimed water, and participating in water trading markets. Despite a recommendation from the National Competition Council (NCC), the NSW government initially refused to allow the company to compete with the state-owned water authority, citing critical public interest issues. However, in 2005 the Australian Competition Tribunal endorsed the recommendation of the NCC with a binding ruling (Australian Competition Tribunal 2005). Accordingly, the NSW government will now need to negotiate an access regime for third parties to Sydney Water sewage infrastructure.

The level of cooperation forthcoming will have a significant impact on the legal and commercial environment in which water authorities in Australia will operate in the future. Until significant precedents have been established, market uncertainties are likely to act as a barrier to long-term investment decisions of both public and private enterprises.

Reform of Complex Governance and Institutional Arrangements

In Australia, water is vested in governments that allow other parties to access and use it for a range of purposes. As noted earlier in this book, however, governance covering the various stages of the water cycle is complex and varies throughout the states. The Senate inquiry into Australia's management of urban water identified this situation as a barrier to achieving greater progress toward more sustainable water management (SECITARC 2002). The inquiry report attributed the source of the barrier to the institutional and policy complexities of three jurisdictions of government, the myriad of agencies responsible for planning, health, environment protection, natural resource management and price regulation, and institutional inertia in general.

The corporate structures via which state governments manage water and sewage also vary considerably. For example, in Melbourne, Sydney, Perth, and Darwin, government-owned corporations manage potable supplies and sewage, but in Adelaide, a privately owned company provides water services under an agreement with the government authority and a public-private multiutility partnership provides services in the ACT.

Many aspects of water management carry significant externalities affecting other areas of water management. The separation of responsibilities means, however, in many cases, that these externalities are not simply assessed (or managed) with due consideration of their wider implications. For example, the assignment of strict pollution limits on sewage discharge may have a major influence on the relative incentives for water reuse.

Water recycling schemes typically aim to integrate various aspects of potable water supply, sewage, and storm water management. Given the existing complexities, some local governments have emphasized the need to establish strong partnerships with state government water authorities to overcome difficulties associated with the division of responsibilities (Chanan and Woods 2005).

Reduced Reliance on Fossil Fuel Consumption and Associated Greenhouse Gas Production

Electricity generation is the largest source of greenhouse gas emissions in Australia, contributing about one third of all emissions in 2002 (Australian Greenhouse Office 2004). This is because most electricity in Australia is produced by the combustion of low cost fossil fuels. Furthermore, in 2004 the Stationary Energy subsector, of which electricity production comprises about 70 percent, was the fastest growing greenhouse gas producer with an expected 43 percent increase on the 1990 level by 2010.

For overall greenhouse gas emissions to be significantly reduced in Australia, their rapid increase from electricity generation needs to be drastically curtailed and eventually reversed. Changes in national water management practices that include increased or new energy-intensive operations would place significant burdens on the national greenhouse gas targets. Unless alternative, nongreenhouse gas producing sources of energy can be implemented, low energy water management solutions will be necessary, which will effectively preclude some technological solutions.

Comprehensive National Guidelines

National guidelines for water recycling are being developed under the auspices of the National Water Quality Management Strategy. The first phase of the guidelines was finalized during 2006. These guidelines have adopted the risk management framework approach pioneered in the 2004 revision of the Australian Drinking Water Guidelines. They cover recycling of treated sewage effluent, grey water, and storm water for nondrinking purposes such as agricultural uses, garden watering, car washing, and toilet flushing. Phase 2 of the National Guidelines includes measures for the implementation and management of potable water recycling schemes.

National guidelines based on appropriate risk and system management criteria have been urgently required in Australia to complement existing guidelines on quality requirement for particular applications. The absence of satisfactory national guidelines has been a hindrance to long-term planning and development of reuse schemes. The lack of definite system management criteria and the singular focus on water quality parameters has been a deterrent to many potential users and suppliers of recycled water, cautious of legal implications.

Conclusion

Technology-based approaches to water management, such as water recycling and desalination, have significant scope for rapid growth in Australia. The necessity to implement such schemes to address freshwater shortages and environmental degradation associated with wastewater discharge is clear. Furthermore, there is significant evidence that Australian authorities now recognize these needs and are beginning to implement policies and actions aimed at encouraging alternative water management strategies.

Although this chapter has specifically identified the need for technological approaches to water management, it is clear that for the most part it is not the technology that is lacking. Suitable water treatment technology for most applications is already in existence and is constantly undergoing improvements in terms of effectiveness and efficiency. To a much greater extent, improvements and reforms are required in social and economic realms. Changes which would facilitate improved technological water management have been identified in terms of institutional structures; government policy; accounting methods; and levels of understanding and communication with stakeholders. This observation serves to drive home the point that water is so fundamentally central to our existence, that no aspects of its management can be legitimately considered in isolation to all others.

Notes

1. This comparison does not consider the question of whether consumers may, in fact, be prepared to pay more for a desalination plant in order to avoid (or lessen) water restrictions; such considerations are more appropriately framed against the background provided by Edwards in Chapter 10.

References

Australian Competition Tribunal. 2005. Application by Services Sydney Pty Limited [2005] ACompT 7. http://www.austlii.edu.au/au/cases/cth/ACompT/2005/7.html (accessed: September 16, 2006).

Australian DEH (Department of the Environment and Heritage). 2006. National Pollutant Inventory: 2004–2005 facility data. http://www.npi.gov.au/ (accessed September 16, 2006).

Australian Greenhouse Office. 2004. *Tracking to the Kyoto target: Australia's greenhouse emissions trends 1990 to 2008–2012 and 2020*. Canberra: Department of the Environment and Heritage.

Chanan, A. and P. Woods. 2005. Managing the water cycle in Sydney metropolitan: Local governments do matter! In *Integrated concepts in water recycling*, edited by S. J. Khan, M. H. Muston, and A. I. Schäfer. Wollongong, NSW: University of Wollongong, 101–107

CoAG (Council of Australian Governments). 2004. *Intergovernmental Agreement on a National Water Initiative, The Commonwealth of Australia and the Governments of New South Wales, Victoria, Queensland, South Australia, the Australian Capital Territory and the Northern Territory*. Canberra: Council of Australian Governments.

Feitz, A. J., O. Braga, G. A. Smythe, and A. I. Schäfer. 2005. Steroid estrogens in marine sediments at proximity to a sewage ocean outfall. In *Integrated concepts in water recycling,* edited by S. J. Khan, M. H. Muston, and A. I. Schäfer, 203–211. Wollongong, NSW: University of Wollongong.

GESAMP (Group of Experts on the Scientific Aspects of Marine Environmental Protection). 1998. Report of the Twenty-Eighth Session of GESAMP. Annex X: GESAMP Statement of 1998 Concerning Marine Pollution Problems, GESAMP Reports and Studies No. 66, IMO/FAO/UNESCO-IOC/WMO/WHO/IAEA/UN/UNEP. Geneva: Joint Group of Experts on the Scientific Aspects of Marine Environmental Protection.

Hamilton, A. J., A-M. Boland, D. Stevens, J. Kelly, J. Radcliffe, A. Ziehrl, P. Dillon, and B. Paulin. 2005. Position of the Australian horticultural industry with respect to the use of reclaimed water. *Agricultural Water Management* 71(3): 181–209.

Hatton MacDonald, D. 2004. *The economics of water: Taking full account of first use, reuse and return to the environment.* Canberra: CSIRO Land and Water.

Hatton MacDonald, D., and B. Dyack. 2004. *Exploring the institutional impediments to conservation and water reuse: National issues.* Canberra: CSIRO Land and Water.

IPART (Independent Pricing and Regulatory Tribunal of New South Wales). 2005. *Investigation into water and wastewater service provision in the Greater Sydney region: Final Report.* Sydney: Independent Pricing and Regulatory Tribunal of New South Wales.

Khan, S. J., A. I Schäfer, and P. Sherman. 2004. Impediments to municipal water recycling in Australia. *Water* 31(2): 114–124.

Marks, J., N. Cromar, H. Fallowfield, and D. Oemcke. 2003. Community experience and perceptions of water reuse. *Water Supply* 3(3): 9–16.

McKay, J., and A. Hurlimann. 2003. Attitudes to reclaimed water for domestic use: Part 1. Age. *Water* 30(5): 45–49.

NRMMC (National Resource Management Ministerial Council). 2005. *National guidelines for water recycling: Managing health and environmental risks.* Draft for Public Consultation. Canberra: NRMMC and Environment Protection and Heritage Council.

N.S.W (New South Wales). DIPNR (Department of Infrastructure Planning and Natural Resources). 2004. *Meeting the challenges—Securing Sydney's water future—The metropolitan water plan.* Sydney: Department of Infrastructure Planning and Natural Resources.

Po, M., J. Kaercher, and B. E. Nancarrow. 2004. *Literature review of factors influencing public perceptions of water reuse.* Canberra: CSIRO Land and Water.

Radcliffe, J. C. 2004. *Water recycling in Australia.* Parkville, Victoria: Australian Academy of Technological Sciences and Engineering.

Rathjen, D., P. Cullen, N. Ashbolt, D. Cunliffe, J. Langford, A. Listowski, J. McKay, T. Priestley, and J. Radcliffe. 2003. *Recycling water for our cities.* Canberra: Prime Minister's Science, Engineering and Innovation Council.

Raucher, R. S. 2005. An economic framework for evaluating water reuse and desal projects: What are the benefits and do they exceed the cost? Paper presented at Water Reuse and Desalination: Mile High Opportunities, 20th Annual WateReuse Symposium. September 2005, Denver, CO.

SECITARC (Senate Environment Communications Information Technology and the Arts References Committee). 2002. *The value of water: Inquiry into Australia's management of urban water.* Canberra: The Parliament of the Commonwealth of Australia.

Sydney Water. 2005. Sydney Water Mandatory Water Restrictions. http://www.sydneywater.com.au/SavingWater/WaterRestrictions/ (accessed December 19, 2005).

United Firefighters Union of Australia. 2005. Recycled Water. *CODE 2: The Official Newsletter of The United Firefighters Union of Australia, Union of Employees, Queensland,* 19(24).

Victoria. DSE (Department of Sustainability and Environment). 2004. *Our Water Our Future—Securing Our Water Future Together.* White Paper. Melbourne: Department of Sustainability and Environment.

Western Australia Government. 2004. *Securing our water future: A state water strategy for western Australia—One year progress report.* Perth: Government of Western Australia.

WSAA (Water Services Association of Australia). 2004. *Health risk assessment of fire fighting from recycled water mains.* Occasional Paper No. 11—November 2004.

WSAA (Water Services Association of Australia). 2005. *Pricing for recycled water.* Occasional Paper No. 12—February 2005.

CHAPTER 13

Water Trading and Market Design

John Rolfe

A KEY DRIVER OF THE WATER REFORM PROCESS in Australia has been the stated desire to secure improvements in economic efficiency. The development of water trading has been one of the more visible aspects of the water reform process. The transfer of water from low value users to high value users should, almost by definition, increase net productivity from units of water. This would lead to gains in economic efficiency and subsequent improvements in social welfare (Freebairn 2005; Young 2005).

In practice, it has turned out to be more difficult to design water markets than was initially expected (Young and McColl 2004). This is in part because of the level of complexity and diversity, which, combined with the variation in external impacts, has made it difficult to develop consistent property rights and trading rules. Australia also has a mature water sector with little opportunity to increase supply to satisfy competing users or resolve conflicts (Young and McColl 2004), and a number of structural and hydrological issues that limit market participation (thin markets). A further issue for market design is that problems of externalities create debates about how property rights are attenuated. Some aspects of the attenuation of rights and the role of spill-over effects have been considered in preceding chapters (see Crase and Dollery, Chapter 6; Crase and O'Keefe, Chapter 11). Overall, this complexity helps to explain why the reforms are not as advanced as in other sectors such as transport or energy, even though Australia has made substantial reforms to improve water access entitlements and adopt pricing structures that reflect true costs (National Competition Council 2004).

The purpose of this chapter is to outline some of the economic principles behind the widely accepted efficiency arguments that attend water trading and to identify opportunities for improving incentive mechanisms. The chapter is structured in the following way. In the next section, some of the issues related to Australian water markets are reviewed. The analysis continues by providing some theoretical background and exploring the benefits and costs associated with water trading. This is followed by a discussion of pertinent market failures before considering the potential contribution of resource economics to market design. Some brief concluding remarks are then provided in the final section.

The Development of Water Markets in Australia

Substantial progress has been made in Australia in the development of water markets (Bennett 2005; Brennan 2006). Trading mechanisms have been established in many catchments, particularly for seasonal allocations (known as temporary transfers). There has also been substantial progress toward the unbundling of land and water titles and the development of property rights systems over water. This has facilitated the trade of water entitlements (known as permanent transfers). As well, there have been improvements in the way the operational costs of maintaining services and delivering water to customers are reflected in annual water charges.

The reform process has established water trading in a number of catchments in Australia, particularly those within the Murray-Darling system where there are large numbers of agricultural water users and a number of supply systems. Trade has been facilitated by the emergence of water agents and the development of public water exchanges such as Watermove (Brennan 2006). Although the system of entitlements varies between states and catchments, the most common is some form of volumetric entitlement that is adjusted proportionally to the availability of supply. Many systems allow two classes of water entitlements, high and medium security, with the former accounting for about 10 percent of entitlements in NSW and most of Victoria (Heaney et al. 2006).

However, trading tends to be restricted within sections of each catchment and within agriculture, essentially restricting the market to a number of sub-sectors. The limited development of the market reflects hydrological, political, and infrastructure barriers. Most of the trading rules are tied to hydrological constraints, reflecting the physical limitations to transferring water (particularly upstream) and carrying it across seasons (Brennan 2006). There has also been sensitivity to the potential transfer of water from agriculture to urban sectors which has largely been addressed by isolating urban demand from potential agricultural suppliers. Bulk supplies to many regional communities have been allocated by regulation (Byrnes et al. 2006), while water shortages for most urban communities have been largely addressed by restrictions on use. Although there is potential for water to be transferred to some urban centers, the lack of suitable (public) infrastructure to transport the water, accompanied by other political constraints, means these markets remain largely undeveloped.

Almost all the activity in water markets has been focused on temporary trades (Brennan 2006). The primary reason for this is that property rights were grandfathered to users who already held licenses or permits under previous systems, thus avoiding the immediate need for market transactions. There is ongoing debate about the efficiency of a trading system dominated by temporary trades. Freebairn and Quiggin (2006) suggest that temporary trades are likely to be associated with larger transaction costs, and thus will be less efficient than permanent trades over the longer term. Alternatively, Brennan (2006) suggests that the visible transaction costs of temporary water trades in the Golbourn-Murray region are very low, and that trade is driven by seasonal variability rather than a desire to achieve a reassignment between productive uses.

Background Economic Theory

Water may be used for a number of different purposes and in different enterprises, which means that large variations can exist in the opportunity costs and benefits of using water. The task is to ensure that the marginal social benefit from each allocation is equivalent (Freebairn 2005). Where the returns are different, then efficiency dividends can be generated by allowing water to be transferred to its highest value use, particularly when these vary in dynamic settings (Easter et al. 1998; Tsur et al. 2004). The development of efficient markets for water has long been advanced as a mechanism to address this allocation problem, both in Australia and overseas (Easter et al. 1998; Bennett 2005). Markets play key signaling roles, where price is used to transfer information about relative scarcity between buyers and sellers. Price also helps to signal the potential desirability of technological improvements and conservation measures.

There are cases where some impacts of water use are not reflected in market prices, meaning that the marginal private benefits of water use differ from the marginal social benefits. This occurs, for example, when there are spillover effects such as salinity or biodiversity losses, and these are not captured within existing allocations of property rights or regulations about water usage. There are some instances where the transfer of water can lead to increased spillover effects, for example where more irrigation supplies are transferred to high salinity risk areas, such as was the case with the pilot interstate water trading project discussed by Crase and Dollery (Chapter 6). In these cases the real problem is that different government interventions may be needed in market allocation systems to control these impacts.

There are three broad challenges facing economists who deal with water allocation issues. The first is the evaluation problem, where the challenge is to identify how optimal an existing level of extraction and allocation of water supplies might be. One difficulty is that water prices tend to be very inaccurate measures of real economic value, as many benefits of water use, such as recreation activities and biodiversity impacts, may not be reflected in market prices (Briscoe 2005). Nevertheless, economists have developed a range of analytical tools to help in the assessment of different impacts (see, for instance, Tsur et al. 2004; Rolfe 2004).

The second major test for economists is in the area of improving the allocation process. It is not always easy to design and initiate market trading systems, especially when property rights may need to be established (or reallocated), where private property rights have to coexist with common property rights, or where substantial uncertainty exists about the extent and pattern of resource use (Quiggin 2001).

The third broad challenge, which is interwoven with the second, is to adjust property rights and/or impose constraints on trading systems to limit potential market failure problems and account for sustainability. This decision process can be viewed as a nested model, where the first choice is between a regulatory and a property rights approach, and the second choice is between flexible and fixed property rights. The first level of choice can be characterized as being between a Pigovian externality framework and a Coasian property rights framework (Anderson 2004), with the former implying much higher levels of direct government control. Should a property rights approach be selected, there are two general philosophies to then establishing them: flexible allocations in the Coase-Posner tradition, and fixed allocations in the Coase-Buchanan tradition (Quiggin 2001). Although flexible allocations have merit when there is variability and uncertainty in supply, fixed allocations have strengths in that they minimize rent-seeking behavior.

Categorization of Benefits from Water Trading

Following the analysis provided by Rolfe (2005) which considered the potential gains from trade within and between irrigated agricultural sectors in northern Australia, the key benefits anticipated in moving from a fixed government system of water allocations to a flexible water trading market in Australia can be categorized along five different lines.

The first are the benefits available from shifting water between sectors. Examples of this are the transfer of water from low value agriculture to high value urban or industrial uses. Much of the recent interest in water trading has focused on the potential gains from intersector trade, although, as noted elsewhere in this volume, a number of impediments to intersectoral transfers remain (see, Edwards, Chapter 10, for a complete discussion).

The second key group of benefits relates to trade within sectors, particularly within agriculture. This occurs when heterogeneity in climate, resources, skills, infrastructure, and other factors means that the opportunity costs of using water varies between farmers. Examples of transfers occurring in Australia include water shifted from low value flood irrigated pasture systems to high value drip irrigated systems for vineyards (Young and McColl 2004). Rolfe (2004) also reports high levels of variation of opportunity costs for water within agricultural sectors in Australia, and Brennan (2006) establishes the importance to farmers of the temporary water market to manage seasonal risks and conditions. The extent of temporary trades in existing water markets in Australia stands as testament to the gains available from reallocating water within the agricultural sector.

The third category of benefit hinges on the impacts on hidden subsidies and the benefits of more accurate pricing that would normally accompany the opening of water markets to competitive forces. Higher water prices would emerge to reflect supply constraints and the status of demand. In turn, this generates benefits for all users insomuch as price signals the benefits of efficient use, and inefficient or low value users tend to exit. Markets also give the latter group a positive incentive to change behavior, and reward them for reducing inefficient practices and/or selling water to higher value users.

The fourth key group of benefits arising from a water market relate to its potential to limit government failure. Arguably, this occurs in two main ways. First, competitive market allocation mechanisms reduce the need for government involvement in allocation choices, and constrain opportunities for governments to be captured by rent seeking. There are additional benefits in terms of conflict resolution insomuch as markets rather than governments deal with competing interests. Second, market mechanisms provide clear signals about resource scarcity and the potential value of providing additional supply. These market signals help to improve the efficiency of infrastructure planning and avoid the types of government failures highlighted by Davidson (1965).

The fifth category of benefits is less tangible, but nonetheless important. They relate to the benefits gained from fostering self-reliance and entrepreneurship in the agricultural sector. The introduction of water trading should encourage greater efficiencies in water use, develop a better skill base for dealing with resources, and make farmers more responsive to changes in factor prices. In this way, a very important benefit to flow from water trading may be a more self-reliant and innovative irrigation sector.

Categorization of Costs from Water Trading

The primary costs that attend a water trade relate to the value of production foregone when water is transferred from a lower value to a higher value use. This is the opportunity cost of water, and, in a well-functioning market, is reflected in the price. There are a number of other costs that may also be associated with the operation of water markets, and these should also be considered in the analysis of potential efficiency gains from market trading and in the design of market mechanisms.

The first group of additional costs to consider is transaction costs. These may be thought of as the search and information costs of finding and negotiating new trading opportunities, and the agreement and enforcement costs of completing transactions (Coggan et al 2005; Freebairn and Quiggan 2006).

Transaction costs tend to be higher when markets are first established, for a number of reasons (Archibald and Renwich 1998; Freebairn and Quiggin 2006). Market prices may not be efficient at signaling demands, there may not be many institutions, or intermediaries that can match potential sellers with buyers, and the initial costs of learning new systems and complying with regulation may be very high. As markets become formalized and participants become more familiar with a system, it is expected that transaction costs will fall. High transaction costs can limit market operation, because buyers and sellers will consider these as well as opportunity costs when evaluating choices. A key focus in the design of property rights and trading mechanisms is to determine which systems generate the lowest transaction costs (Freebairn and Quiggin 2006; Young and McColl 2004).

The second group of additional costs to consider is transition costs. These are essentially the costs involved in changing property rights and institutional systems to suit the development of water markets. Transition costs can be important, particularly when changes are associated with uncertainty and/or irreversibility. Quiggin (2001) describes how property rights for water can be established in two broad frameworks. The flexible rights approach, characterized as a share of the available water resource, has advantages in that rights can be subsequently adjusted to account for changing conditions or new information. This may be very important for a transitory and complex resource such as water, where supplies are affected by factors such as changing weather patterns and knowledge might be gained about system losses and groundwater interactions.

However, there are two main difficulties identified with the flexible rights approach. The first is that it provides incentives for continued rent-seeking behavior, as stakeholders try to have rights redefined in their favor (Quiggin 2001). The second is that flexible rights create uncertainty about the security of the water titles, which can be an inhibitor to trade (Archibald and Renwick 1998). For these reasons, the establishment of property rights over water has focused on fixed rights rather than flexible rights (Quiggin 2001). Fixed rights have advantages in that they create certainty about the bundle of goods on offer, thus facilitating trade. However, fixed rights may be associated with higher transition costs.

There are three particular types of transition costs associated with establishing water markets and fixed water rights. The first relates to instances where water is over-allocated. This can occur when licenses are initially over-allocated and then converted to a stronger form of property right, or when subsequent information reveals that water supplies (e.g., groundwater) were less extensive than earlier estimates. There have been some commitments of public funds in Australia to addressing these over-allocation issues because the fixed property rights make it difficult to reallocate (nonexistent) supplies to agriculture without compensation. Some of these problems have been minimized by giving fixed property rights over shares in water resources (as detailed in the National Water Initiative) rather than establishing fixed volumetric allocations.

The second type of transition cost associated with establishing water markets occurs when property rights allocations do not match well with the stocks and flows of water in the hydrological cycle (Coggan et al. 2005). A core issue in this context is the property rights to recharge water (Young and McColl 2006). In many irrigated agricultural systems, substantial amounts of water diverted for agriculture are not used by plants, but enter groundwater systems or return to streams. This means that many systems of fixed property rights can involve the same units of water, which returns to a river system after being applied in irrigation (Coggan et al. 2005). Difficulties arise when irrigation systems change, or when water is diverted to other uses, and the level of subsequent recharges is diminished. These affect remaining property rights (or environmental allocations), and are often a hidden transition cost of improving efficiency signals in water markets. The problem of ill-defined return flows from irrigation was discussed in earlier chapters (Crase and Dollery, Chapter 6) and remains a major challenge for policymakers.

The third type of transition cost associated with establishing water markets relates to the costs of designing and changing institutions. Institutions are very important for providing the framework for achieving change and dealing with political economy issues, and there is substantial focus in water research on the types of institutions that are most efficient at generating change and minimizing transaction costs (e.g., Challen 2000). There are also a range of institutional mechanisms for dealing with the externality and public good impacts of water allocations. The establishment of water markets and property rights often impacts on the design of these institutions, or necessitates the creation of new ones. In both cases there may be substantial costs involved.

A final set of additional costs to consider is the administration costs of government. The system of market exchange and property rights needs to be administered and enforced to be effective. There will be establishment costs and ongoing administration costs to be borne by government, although administrative costs of a market system are likely to be lower than the administrative costs of a government-controlled system.

Potential Market Failure Issues

The discussion of costs and benefits has treated water as a single commodity, albeit with reference to the hydrological cycle from which water is drawn. This treatment is simplistic, and any discussion of the development of water markets and trading issues has to extend to a consideration of the different interactions that might compromise the potential efficiency of a market framework. Drawing on economic theory, it is possible to identify a number of areas where complex interactions may limit the efficiency of a market framework.

The first instance occurs where market operations cause impacts on other parties. This may include downstream impacts of reduced water flows, poorer water quality, and increased salinisation, as well as impacts on biodiversity, and social impacts, such as the problems of stranded assets. Under the Pigovian approach, these occurrences are treated as negative externalities, where the spillover costs of actions are forced onto nonconsenting third parties.

The standard solution to externality problems is to signal the real costs of their actions to the parties causing the spillover. In practice, this typically involves the imposition of pollution taxes or congestion taxes, or the use of regulations to control and limit the actions causing negative spillovers. Both alternatives require substantial government involvement, which may be difficult to reconcile in the establishment of a water trading system where market forces interact to solve allocation problems. There may also be equity issues involved in the use of taxes or regulations to control external impacts after investment decisions have been made (Quiggin 2001), and there may be a number of efficiency issues in using relatively blunt and stable instruments such as taxes to deal with complex and varying external impacts.

The other main approach to dealing with external impacts is the Coasian property rights approach (Anderson 2004). Here the rights of both internal and external parties may be viewed as competing interests in a resource, where the challenge is to establish property rights so that parties can then negotiate acceptable solutions. In the absence of transaction costs, this is a more efficient approach than the Pigovian framework. The property rights approach sits comfortably with a water market setting, as parties with specific rights would be able to trade to find optimal solutions; however, when the transaction costs of specification and engagement are considered, a property rights approach will not always be efficient. Transaction costs would be expected to rise across multiple impacts and multiple parties, and where impacts occur across time and space. The key challenge is to specify entitlements when these complexities are involved.

Goods that share nonrival and/or nonexclusive characteristics are difficult to provide through market processes. There is no private incentive for markets to supply goods that have these characteristics, because the benefits can be enjoyed by more than one person simultaneously, and there is no economic way of excluding noncontributors. For example, it is difficult for private markets to supply mitigation actions such as improvements in water quality because while the costs of mitigation might accrue to the individual or the enterprise, the benefits are often much more widely spread and beneficiaries will tend to free ride on the services provided.

The economic characteristics of water demand and supply range across a continuum of rival to non-rival aspects (Young 2005). Water that is diverted for private use is akin to a rival good, but operations that cause impacts on other parties share more nonrival characteristics. In a similar way the exclusiveness of water benefits and associated impacts varies from being relatively easy to exclude to almost nonexcludable. Aspects of water use that have nonrival and nonexclusive characteristics are akin to public goods and need government involvement to address supply issues.

Salinity and water quality issues illustrate a number of nonrival and nonexclusive characteristics. Some salinity and water quality impacts may occur on-site, rather than external to a farm. Where the impacts are on future production, these should be more properly classed as sustainability issues (Pannell et al. 2001). However, other salinity and water quality impacts may occur off-farm, perhaps over long distances and with long time lags. There are typically many contributors and many affected parties with these types of impacts, and it is usually impossible to link a particular impact with a specific contributor. The impacts share the characteristics of nonrival and nonexclusive bads, and remedies to these problems often share the characteristics of nonrival and nonexclusive goods. In many cases the remedies may provide a mixture of rival goods (e.g., the rehabilitation of land on a specific farm) as well as nonrival goods (improvement in biodiversity protection).

The third limitation on the operation of the market relates to notions of sustainability. Although the number of different concepts of sustainability makes it hard to define accurately (as noted from a legal perspective by McKay in Chapter 4), there seems to be a general understanding that some stock of natural resources should be maintained for future generations (Quiggin 2001). The implication is that problems such as salinity, rising water tables, sediment discharge, and water extraction levels should not exceed certain thresholds. Perhaps the most visible manifestation of the sustainability arguments are the minimum standards adopted for environmental flows in most regulated systems, and the consistent pressure for environmental flows to be increased (Freebairn 2005). These demands are also evidence of the policymakers' concerns with the quantity of flow rather than the timing of flows and the nexus with environmental outcomes. Although sustainability is often difficult to theoretically ground (Quiggin 2001), the practical application of sustainability demands through a political economy context often imposes the limits within which water markets can operate.

These different considerations of external impacts, nonrival and nonexclusive characteristics, and sustainability goals impact on water markets in several ways. The first is that water markets are often limited to certain bundles of water allocations within defined areas. The use of market forces is normally constrained to a subset of water stocks and flows in the hydrological cycle. These limits to the framing of water markets relates to the treatment of environmental flows. A portion of water stocks and in-stream flows is normally reserved for environmental purposes. In some cases, such as the Snowy River, water has been transferred back into the river for environmental flows. In other cases, such as the Murray, some flows are being pulsed down the river to create flooding events for wetland regeneration and water bird breeding. The complexity of managing these arrangements is highlighted by Hillman in Chapter 9.

The second outcome is that the property rights that may be created over water are often restricted in some ways to minimize offsite impacts. Key restrictions relate to rules about where water can be transferred, and the use of conversion ratios to take account of potential hydrological losses and/or salinity impacts. There are also a number of constraints on water markets and water uses imposed through regulatory mechanisms. These include rules about how water might be accessed (e.g., stream flow thresholds before harvesting is allowed), water use (through property planning requirements in some areas), and water return (limits on the levels of contaminants in the water). Regulation has also extended to capture other interactions with water use, such as restrictions on vegetation clearing, which are partly justified as ways of avoiding dryland salinity and other land degradation problems.

Government influence on water markets is not only concentrated on the setting of market limits, adjustments to property rights, and the use of regulatory mechanisms. In some cases governments can enter water markets as a participant, perhaps to purchase additional environmental flows, although they have generally been reluctant to pursue this approach in most Australian jurisdictions. Governments are also transferring some funding and responsibility for environmental protection and restoration to regional community bodies, and these may interact with water markets in different ways, particularly in relation to encouraging minimum standards in relation to environmental conditions and water reuse. Notwithstanding this progress there is still much room for improvement on the policy front.

Potential Contribution of Resource Economics to Water Market Design

Economic principles and tools can be applied to improve the design of water markets in three broad areas. The first is better identification of nonproductive values of water and evaluation of efficient allocation levels, the second is the design of efficient market processes, while the third is the potential use of market-based mechanisms to address externalities and public good aspects of water. Here, the three broad areas are briefly reviewed with particular focus on recent Australian experiences.

A lack of valuation studies has limited the contribution of economics to the debate over how water should be allocated between the environment and other competing uses. Instead, decisions about the allocation process have been made in other ways. In the Murray-Darling system, where nearly 80 percent of natural flows are diverted for commercial uses (Young and McColl 2004), initial licenses for water use have largely been grandfathered into property rights entitlements, maintaining high levels of extraction. Although there is a commitment under the National Water Initiative to return 500GL of environmental flows to the river, the rational basis for setting this level is not transparent. In river systems with lower levels of development, extractions have tended to be capped at 50 percent of median flow levels, or in the case of the wild rivers in far north Queensland, at zero extractions. These benchmarks have been set through scientific, consultation, and political processes rather than through any formal evaluation of net social benefits. The axing of the Choice Modeling component from the Living Murray analysis is illustrative of this problem and has been detailed in an earlier chapter (Crase and O'Keefe, Chapter 11).

There are a number of studies that can potentially be used to assess the broader question of allocative efficiency of water resources. Recreation valuation studies for wetlands or water resources have been reported by Whitten and Bennett (2002), Rolfe and Prayaga (2006), and Dyack et al. (2006). In addition, Choice Modeling studies are now available that provide some assessment of the values that attend biodiversity impacts relating to wetland protection. These cover a variety of locations including the Macquarie Marshes and Gywder wetlands of New South Wales (Morrison et al. 2002), wetlands protection in New South Wales and South Australia (Whitten and Bennett 2005), and floodplain protection in Queensland (Rolfe and Windle 2005). Morrison and Bennett (2004) also report the development of a value template for healthy rivers in New South Wales that would facilitate the transfer of value estimates for water quality between different locations. In sum, there is ample extant data that could be applied to assist in formulating allocatively efficient policies.

Although economic processes can also be used to adjust the allocation of water resources between environment and consumptive uses, there is little evidence of this happening to date in Australia. The two main options for making adjustments are to send price signals about inefficient allocations and allow the market to make adjustments, or to make water purchases (and sales) on behalf of those representing environmental interests. The first option would require the inclusion of the costs of any external effects, including those on biodiversity, within water prices, while the second might involve the relaxation of set thresholds about environmental use. Currently, the combined lack of detailed economic information and opposition from a variety of interest groups are restricting opportunities for the broader allocation issues to be solved with market instruments.

The second broad area of contribution from economics relates to the design of market processes. Although there is now widespread adoption of trading mechanisms to allocate water within consumptive uses (apart from urban customers), there are a number of ongoing debates about how to improve the trading processes. Research into the structure of water entitlements, property rights, transaction costs, risk and uncertainty are all important in this context (Young and McColl 2004; Freebairn and Quiggin 2006; Brennan 2006). There is also potential for economic theory and market analysis to continue improving the design of water markets, as well as promise for the use of financial market tools such as derivatives. In this regard, options contracts that take advantage of the heterogeneity of demand across urban and agricultural users have, to date, been largely dismissed in Australia. This approach has usually been taken on the grounds that Australian water markets lack maturity. Ironically, such approaches have been used abroad to successfully alleviate the tension arising from agricultural-urban water trade and thereby encourage maturation of the market (see, for example, Metropolitan Water District of Southern California 2004).

The third broad area of contribution from economics relates to the design of market mechanisms when there are external impacts and sustainability objectives. Young and McColl (2004) suggest that it may be more efficient to separate the management of environmental impacts from the definition of water allocations and entitlements. If environmental impacts were regulated separately, it would simplify the identification and trading of entitlements, leading to a reduction in transaction costs. An alternative to a reliance on regulation to deal with environmental impacts is to adopt market based incentives for this purpose.[1]

Quantity control mechanisms offer some promise for limiting negative outcomes of water usage while retaining the flexibility and benefits of market processes. The Hunter River Salinity Trading Scheme operates in New South Wales to manage saline water discharge from mostly industrial sites (coal mines). There is potential to extend similar mechanisms to cover agricultural emitters. O'Dea and Rolfe (2005) reviewed the use of cap-and-trade mechanisms to deal with water quality issues. They found that there were several requirements for success: there needs to be a core of point source emitters involved; strong political will and regulatory mechanisms to set a cap are required; and good scientific knowledge, measurement, and monitoring systems must be available. Given the limited state of our knowledge and the fact that irrigation accounts for most water use and is spread across thousands of farmers and vast geographical areas, this approach appears of limited potential in the immediate term. Nevertheless, further developments in this area over time may see water markets integrated with emission trading mechanisms to minimize off-site and downstream impacts of water use.

Conclusion

In line with a number of international trends, the water reform process in Australia has involved moves to treating water as an economic asset rather than as a subsidized political good, although the transformation remains incomplete. The changes have seen moves to price water more efficiently, as well as to allow market forces to allocate water between competing users. Such an allocation process has clear economic benefits when water can be transferred from low value use to high value use. Although there has been a great deal of focus on the institutional changes necessary to achieve these outcomes, there has been less focus on the economic gains from establishing water markets and on the potential for economic processes to solve wider allocation and environmental impacts issues. This may be because of the complexity of water issues in localized areas, the limited economic data that is currently available, and also because of the lack of specialist skills needed to perform the economic analysis.

Whereas the benefits of moving to competitive market structures for water may be greater than is often recognized, there may also be major costs to consider. Apart from the opportunity costs (which will be easier to measure with better functioning markets and the use of specialized valuation techniques), there may be some higher transaction costs and initial transition costs to consider. Those potential costs rise with higher levels of risk and uncertainty about the stocks and flows of water resources, because it can be difficult to adjust property rights and market structures should better information become available at a later date.

Establishing markets over water is challenging because of the interactions involved in the hydrological cycle and the potential for off-site impacts and non-rival effects to attend market transactions. These factors make both the analysis of potential efficiency gains and the design of better allocation mechanisms complex processes. The development of more sophisticated valuation approaches for both the direct and indirect impacts of different allocations is improving insights into the real net gains available from water transfers. However, more studies are needed in Australian contexts.

A variety of approaches are being used to improve allocation of water resources. Markets are typically established over only a subset of water stocks and flows, with nonmarket reserves used to meet environmental and sustainability goals, and regulatory processes used to deal with urban allocations. Property rights are being designed so as to limit the variety of possible third-party impacts, and there are still a variety of regulatory mechanisms employed. The current interest in trialing the use of market based instruments in Australia to deal with problems like salinity and water quality suggest there is potential for competitive market processes to not only allocate water resources efficiently between competing users, but also to allocate effort relating to mitigation and control of negative impacts associated with water use. The key challenges in designing market mechanisms are presently twofold: to expand current markets to cover most consumptive uses and to design mechanisms to deal with broader allocation and environmental impacts issues. This volume provides practical insights into both of these challenges, although much additional policy work is required.

Notes

1. For a description of the types of instruments used in this context see, for instance, Sterner 2003.

References

Anderson, T. 2004. Donning Coase-coloured glasses: A property rights view of natural resource economics. *Australian Journal of Agricultural and Resource Economics* 48(3): 445–462.

Archibald, S. O., and M. E. Renwick. 1998. Expected transaction costs and incentives for water market development. In *Markets for water: Potential and performance*, edited by K. W. Easter, M. W. Rosegrant, and A. Dinar, 95–117. London: Kluwer Academic Publishers.

Bennett, J., ed. 2005. *The evolution of markets for water: Theory and practice in Australia.* Cheltehman, U.K.: Edward Elgar.

Brennan, D. 2006. Water policy reform in Australia: Lessons from the Victorian seasonal water market. *Australian Journal of Agricultural and Resource Economics* 50:403–423.

Briscoe, J. 2005. Water as an economic good. In *Cost-benefit analysis and water resources management*, edited by R. Brouwer and D. Pearce, 46–70. Cheltenham: Edward Elgar.

Byrnes, J., L. Crase, and B. Dollery. 2006. Regulation versus pricing in urban water policy: The case of the Australian National Water Initiative. *Australian Journal of Agricultural and Resource Economics* 50: 437–449.

Challen, R. 2000. *Institutions, transactions costs and environmental policy: Institutional reform for water policy.* Cheltenham, U.K.: Edward Elgar.

Coggan, A., S. Whitten, and N. Abel. 2005. Accounting for water flows: Are entitlements to water complete and defensible and does this matter? In *The evolution of markets for water: theory and practice in Australia*, edited by J. Bennett, 94–118. Cheltenham, U.K.: Edward Elgar.

Davidson, B. 1965. *The northern myth: A study of the physical and economic limits to agricultural and pastoral development in northern Australia.* Melbourne: Melbourne University Press.

Dyack, B., J. Rolfe, J. Harvey, D. O'Connell, and N. Abel. 2006. *Recreation Values: Coorong and Barmah Forest Case Studies.* CSIRO: Water for a Healthy Country National Research Flagship, Canberra.

Easter, K. W., M. W. Rosegrant, and A. Dinar. 1998. Water markets: Transaction costs and institutional options. In *Markets for water: Potential and performance*, edited by K. W. Easter, M. W. Rosegrant, and A. Dinar, 1–18. London: Kluwer Academic Publishers.

Freebairn, J. 2005. Principles and issues for effective Australian water markets. In *The evolution of markets for water: Theory and practice in Australia*, edited by J. Bennett, 8–23. Cheltenham, U.K.: Edward Elgar.

Freebairn, J., and J. Quiggan. 2006. Water rights for variable supplies. *Australian Journal of Agricultural and Resource Economics* 50:295–312.

Heaney, A., G. Dwyer, S. Beare, D. Peterson, and L. Pechey. 2006. Third-party effects of water trading and potential policy responses. *Australian Journal of Agricultural and Resource Economics* 50:277–293.

Metropolitan Water District of Southern California. 2004. *Landmark Ag-to-Urban Water Transfer Further Diversifies and Buttresses Southland's Water Supplies for Coming Decade.* California: MWDSC. http://www.mwdh2o.com/mwdh2o/pages/news/press_releases/2004–05/PaloVerde.htm (accessed 2nd October 2006).

Morrison, M., and J. Bennett. 2004. Valuing New South Wales rivers for use in benefit transfer. *Australian Journal of Agricultural and Resource Economics* 48(4): 591–611.

Morrison, M., J. Bennett, R. Blamey, and J. Louviere. 2002. Choice modelling and tests of benefit transfer. *American Journal of Agricultural Economics* 84(1): 161–170.

National Competition Council. 2004. *NCC occasional series: Microeconomic reform in Australia—Comparison to other OECD countries*, AusInfo, Canberra.

O'Dea, G., and J. Rolfe. 2005. *How Viable are Cap-and-Trade Mechanisms in Addressing Water Quality Issues in the Lower Fitzroy.* Establishing the Potential For Offset Trading In The Lower Fitzroy River. Research Report No 4. Rockhampton, QLD: Central Queensland University.

Ostrom, E. 1990. *Governing the Commons: The evolution of instruments for collective action.* Cambridge: Cambridge University Press.

Pannell, D. J., D. J. McFarlane, and R. Ferdowsian. 2001. Rethinking the externality issue for dryland salinity in Western Australia. *Australian Journal of Agricultural and Resource Economics* 45(3): 459–476.

Quiggin, J. 2001. Environmental economics and the Murray-Darling system. *Australian Journal of Agricultural and Resource Economics* 45(1): 67–94.

Rolfe, J. C. 2004. Assessing Demands for Irrigation Water in North Queensland, *Australasian Agribusiness Review* 12. http://www.agribusiness.asn.au/Publications_Review/Review_Vol12_1/Rolfe.htm (accessed December 10, 2005).

Rolfe, J. 2005. Potential efficiency gains from water trading in Queensland. In *The evolution of markets for water: Theory and practice in Australia*, edited by J. Bennett, 119–138. Cheltenham, U.K.: Edward Elgar.

Rolfe, J. and Prayaga, P. 2006. Estimating values for recreational fishing at freshwater dams in Queensland. *Australian Journal of Agricultural and Resource Economics*, in press.

Rolfe, J., and J. Windle. 2005. Option values for reserve water in the Fitzroy Basin. *Australian Journal of Agricultural and Resource Economics* 49(1): 91–114.

Sterner, T. 2003. *Policy instruments for environmental and natural resource management.* Washington, DC: Resources for the Future.

Tsur, Y., T. Roe, R. Doukkali, and A. Dinar. 2004. *Pricing irrigation water: Principles and cases from developing countries.* Washington, DC: Resources for the Future.

Whitten, S. M., and J. W. Bennett. 2002. A travel cost study of duck hunting in the upper south east of South Australia. *Australian Geographer* 33(2): 207–221.

Whitten, S., and J. Bennett. 2005. *Managing wetlands for private and social good.* Cheltenham: Edward Elgar.

Young, M. D., and J.C. McColl. 2004. The right to water: "Ownership" and responsibility. *Dialogue* 23(3): 4–17.

Young, R. A. 2005. Economic criteria for water allocation and valuation. In *Cost-benefit analysis and water resources management*, edited by R. Brouwer, and D. Pearce, 13–45. Cheltenham, U.K.: Edward Elgar.

CHAPTER 14

Adaptive Management

Phil Pagan

*I*N CHAPTER 5, JOHN QUIGGIN DREW an important distinction between the concepts of risk and uncertainty and also outlined broad principles for dealing with both in water management. This chapter focuses on the use of adaptive management principles as a way of coping with the pervasive uncertainty that characterizes the Australian water management landscape.

This chapter is organized around the following questions:

- What is adaptive management?[1]
- What is the justification for an adaptive approach to water management in the Australian milieu?[1]
- What are the past and current experiences in adaptation of Australian water management arrangements?
- What are the opportunities and barriers to the future incorporation of adaptive management principles?

The chapter focuses primarily on Australian data and experiences, giving special attention to the role of adaptive management in the National Water Initiative and the implications of "The Living Murray" process.

What Is Adaptive Management?

Adaptive management in a general sense involves learning from management actions and using that learning to improve the next stage of management (Lessard 1998; Allan and Curtis 2003a). Adaptive management concepts were first proposed in relation to environmental management in the 1970s (e.g., Holling 1978). It has since been accepted as a valuable approach by natural resource managers, both internationally and throughout Australia. Moreover, adaptive management principles are now recognized at broad policy and legislative levels and are inherent within management strategies implemented at a range of spatial scales across Australia (national, state, catchment, and subcatchment), for a range of natural resources, including water. Examples specific to water include:

- The NSW Water Management Act 2000 explicitly requires that adaptive management principles are followed in the management of NSW water resources.
- A key outcome of the water access and entitlements framework of the National Water Initiative is to provide for adaptive management of surface and groundwater systems (Council of Australian Governments 2004a).
- The Living Murray Business Plan, which outlines implementation of the *Intergovernmental Agreement on Addressing Water Overallocation and Achieving Environmental Objectives in the Murray Darling Basin,* requires an adaptive management approach to be adopted (MDBC 2005).

The original articulation of adaptive management by Holling (1978) and Walters (1986) refers to the application of experimentation to the design and implementation of natural resource and environmental management policies. McLain and Lee (1996) note that the idea of adaptive management was itself derived from adaptive control process theory, which was developed to address technical issues involved in the construction of decisionmaking devices, or control devices, capable of learning from experience. The key attribute contributed by adaptive control process theory was the concept of incorporating feedback loops into the management process in order to accelerate the rate at which environmental decisionmakers learn from experience. Adaptive management encourages managers to structure interventions in ways that permit them to anticipate as a tool for learning, rather than attempting to avoid or merely react to inevitable surprises.

Notwithstanding this common core, adaptive management is interpreted in different ways. McLain and Lee (1996) observe alternative views, such as those proffered by local natural resource managers who will often interpret adaptive management to mean the development of predictive tools that can be used for site-specific management. Alternatively, policymakers will often interpret the approach as a means of seeing the broad impact of policies, generally through the development of large-scale models.

Alternative Forms of Adaptive Management

There are three ways to structure management as an adaptive process:

- Evolutionary or trial and error, in which early choices are essentially haphazard, whereas later choices are made from a subset that gives better results.
- A passive adaptive approach, where lessons from the past are used to design a single best policy to apply currently. This involves using historical data available at the time to construct a single-best estimate or model for response, and the decision choice is based on assuming that this model is correct.

- Active adaptive management, where policy and its implementation are used as tools for accelerated learning. This involves using data that is available at each time period to structure a range of alternative response models, and a policy choice is made that reflects some computed balance between expected short-term performance and the long-term value of knowing (Walters and Holling 1990; Taylor et al. 1997; Allan and Curtis 2003a).

It is this last interpretation that is generally espoused by proponents of the adaptive management approach. In particular, the original articulation of the Adaptive Environmental Assessment and Management (AEAM) process was firmly based on the active adaptive management philosophy (Holling 1978). A key characteristic of an *active* adaptive management regime is that specific hypotheses must be rigorously tested and evaluated so as to provide input into future management practices (Taylor et al. 1997). For this approach to be successful, a controlled experiment must be permitted to run for a period so that the influence of persistent variables can be assessed.

Although the active adaptive management approach may be preferred by advocates, the discussion later in this chapter demonstrates that the weaker forms have been more regularly applied.

What Is the Justification for an Adaptive Approach to Water Management in the Australian Milieu?

The appeal of adaptive management in relation to water resource management is driven by three factors that are common across the management of most natural resources. These are:

- our generally rudimentary understanding of most natural systems (which arises from their complexity);
- the continually changing community goals and expectations that management policies are intended to pursue; and
- the fact that most natural systems don't reach a persistent equilibrium state.

For most natural resource management (NRM) issues, the combination of the above factors eliminates the possibility of developing and implementing feasible and appropriate policy that will not need to be changed over time—hence the need for *adaptive* management.

Two of the above reasons are commonly cited for adopting an adaptive approach to the management of natural resources. First, our level of knowledge about most natural systems is incomplete and a capacity to respond to new knowledge as it becomes available is necessary. In Chapter 11, Crase and O'Keefe canvassed the uncertain and changing nature of our understanding of technical relationships underpinning Australian riverine ecosystems. In addition, not only is it necessary to be able to respond to new information, but also it is generally inefficient to delay action in the hope that full information about natural system characteristics and interactions will become available. Blann and

Light (2000) propose that the prospect of accumulating sufficient knowledge to allow control of natural resource outcomes is an illusion. Second, the objectives and preferences of the community in relation to natural resources are likely to change over time. Accordingly, it is essential to have the capacity to reflect these changing community values within policy objectives (e.g., Productivity Commission 2003).

The third reason noted previously, that natural systems generally do not move to a stable equilibrium state (Holling 1978), is less commonly identified as a reason for taking an adaptive management approach. Holling identified four properties that determine how ecological systems respond to change:

- The parts of an ecological system are connected to each other in a selective way that has implications for what should be measured.
- Events are not uniform over space, which has implications for how intense impacts will be and where they will occur.
- Sharp shifts in behavior are natural for many ecosystems. Traditional methods of monitoring and assessment can misinterpret these and make them seem unexpected or perverse.
- Variability, not constancy, is a feature of ecological systems that contributes to their persistence and to their self-monitoring and self-correcting capacities.

The result of the consistent presence of these characteristics is that ecological systems do not evolve toward some climax or persistent equilibrium, but are dominated by a pattern in time and space of growth, quiescence, collapse, and renewal (Holling 1978). Walters (1986) also observed these characteristics of natural system dynamics, describing them as "a rhythm of crisis and opportunities in resource systems." Importantly, these patterns are also observable in certain social structures such as businesses (Hax and Wilde 1999).

It would be difficult to overstate the significance of this third reason for adopting an adaptive management approach. It could be expected that policy changes as a result of improvements in knowledge about ecosystem function (the first justification), and associated changes in community attitudes (the second justification) will incrementally allow movement to a more informed, stable, and efficient equilibrium state of management for particular ecological systems. However, this third justification highlights the dynamic nature of most ecological systems, and that these systems may naturally move to an alternate state of operation because of influences quite separate to the effects of the human management regime that is in place. Indeed, it should be expected that such changes will inevitably occur, regardless of the quality of management intervention. As a consequence, an adaptive approach to management is essential, not only in the short term, while information failures are rectified through research, but also as a long-term management approach. Put simply, even comprehensive knowledge of current system interactions is unlikely to be sufficient in the future.

Having established the case for an adaptive management philosophy, the next section reviews past applications of adaptive management principles in the management of Australian water resources. Following that, the opportunities and barriers for the improved application of adaptive management principles are considered.

What are the Past and Current Experiences in Adaptation of Australian Water Management Arrangements?

This section reflects on Australian experiences in the adaptive management of water resources. It particularly focuses on past attempts to follow an active adaptive management approach, and considers the potential for active adaptive management within key current Australian water management reform processes— The Living Murray (TLM) program, and the National Water Initiative (NWI).

The Background of Historical Change

As discussed by Musgrave in Chapter 3, there have been a number of distinct phases in water management in Australia over the last 150 years. Phases of water management have included: development under ignorance of the environmental consequences; conscious exploitation (under assumptions of minimal harm); and, more recently, management based on the need to maintain a broad range of (ecosystem) services through sustainable management (Barr and Cary 1992).

These striking changes in the overarching goals of water resources management have manifested in all three justifications for an adaptive approach. That is, there has been a dramatic increase in knowledge about riverine ecosystems and about the role of, and interactions involving, water in the landscape more generally. There have also been significant changes in system dynamics as a response to the regulation of waterways, changing water extraction demands, natural climate variability, and perhaps longer term climatic trends. Finally, there have been dramatic changes in community goals and expectations that water management policies have needed to pursue. For example, consider the historical social objectives that have shaped water management policy in the past. These have included the development of a sturdy yeomanry husbanding the land, and social objectives relating to providing security from northern invaders. Current social objectives which influence Australian water management policy focus on national and regional economic development, but there is recognition that this must be firmly constrained within the ecological capacity of riverine environments. Emerging social objectives which may play a much more significant role in the management of Australian water resources in the future include a clearer recognition of indigenous interests (both cultural and economic).

It seems plausible to classify much of the adaptation in water management policies that has emerged in the past as being of an evolutionary nature. Regularly, the adaptation has occurred following sustained periods of public and interest group objection to obvious and significant degradation, or to other outcomes that are inconsistent with the needs of water users. That is, in most cases adaptation in management has approximated a trial and error approach, in which early choices have been reasonably haphazard, with later choices being made from a subset that have been demonstrated to produce better results. Historical adaptation in water management policy has certainly not been driven by purposeful experimentation, monitoring, and testing of alternative hypotheses about response models that would characterize an active adaptive management approach.

The previously mentioned observation is certainly the case for state-based water reform processes such as has occurred in New South Wales (NSW) in recent times. The current NSW (Labor) government has made water management reform aimed at ensuring socially, economically, and environmentally sustainable water management practices a key component of its environment agenda since being first elected in March 1995.[2] The reform process has covered a wide range of management arrangements, including establishing appropriate water quality and quantity objectives, addressing overallocation of water extraction rights, reforming water pricing and water trading arrangements, restructuring NSW water businesses, and setting ongoing sharing arrangements through the development of various forms of water sharing plans. But both the approach and policy settings to achieve these have continued to change. Major state-wide water reform process changes were made by the NSW government in 1995, 1997, 2001, and 2003. The form and structure of each new approach appears to have been driven as much by the need for a political response to community impressions of progress, as by any objective assessment of performance of the approach being replaced. It is only now, in 2006, that a comprehensive system of monitoring, evaluation, and reporting is being developed for assessment of water sharing plans. In addition, the nature of these water sharing plans is such that they do not encourage the testing of alternative response models.

Past Experiences in the Application of Active Adaptive Management in Australian Water Management

Allan and Curtis (2003b) report that at a workshop of adaptive management practitioners held in Australia in 2002, it was apparent that there were strongly divergent views about the nature, scope, and requirements of an adaptive management approach. This is also apparent when assessing the interpretations of the AEAM, as a specific form of active adaptive management, in the Australian literature. For example the CRC for Catchment Hydrology (2005), while recognizing that it is a process that is broader than modeling, restrict their discussion of AEAM to a specific computer based model framework that can be used to involve participants in priority setting and understanding basic influences on water quality in a catchment. The CRC further contends that these types of AEAM models "can be developed in the order of a few weeks to a couple of months, including testing"(CRC for Catchment Hydrology 2005, 23).

This is clearly not the type of active adaptive management approach that Holling (1978) envisaged when initially devising the AEAM approach. However, the Australian examples of the application of active adaptive management to water and riverine management, which take a view more consistent with Holling's prescriptions, include a strong focus on the initial development and use of computer models to facilitate community involvement in the examination of alternative management scenarios. These include applications in the La Trobe and Goulburn Rivers in Victoria, the North Johnston River catchment in Queensland, and the Macquarie Marshes, South Creek, Tuggerah Lakes, and Sydney Water catchment in NSW.

Gilmour et al. (1999) review the application of adaptive management in the last three of these locations. The three projects (South Creek, Tuggerah Lakes, and Sydney Water catchment) all considered water quality issues in peri-urban catchments around Sydney as their primary focus. All included workshop-based activities, focusing on model development and use, to analyze a range of water management alternatives to address significant and contentious issues in the community. The modeling phase for each application was followed by attempts to implement potential solutions. The evaluation of these processes by Gilmour et al. (1999) makes the following observations:

- The selection of workshop participants was crucial, and it required people who can boldly represent the community, and who have broad community support, while at the same time providing the range of specialist technical skills to ensure that a balanced discussion is maintained and that the systems analysis is accurate and comprehensive.

- The issue of getting agencies to implement AEAM recommendations is more complex than is suggested by questions of self-interest and self-protection. In all three cases, the agencies were struggling with the political overtones of environmental management problems. They were all seeking a technical solution with broad community support, and chose the AEAM approach as a potentially successful path. However, none of the sponsors appeared to have thought through the issues involved in implementing the workshop outcomes. Nevertheless, an adaptive management strategy was subsequently implemented for the Tuggerah Lakes and was regarded as being a viable option for further progressing management of Sydney's water supply catchment.

- It was clear from the South Creek and Tuggerah Lakes projects that there are important long-term benefits to be gained from developing community support for adaptive management strategies. Ecosystem management is a long-term process. In both cases, a single agency has had the lead role in driving follow-through, even though there were multiple organizations involved in the workshops. Ironically, events in the lead agencies in both projects demonstrate the vulnerability of adaptive management strategies to organizational change.

- The reviewers noted that in all three of the projects, political pressures to create broadly acceptable solutions made a participatory process attractive to the lead institutions. Yet, having achieved a degree of consensus, with a resulting reduction in political pressure, each lead institution scaled back the level of community involvement. They observed a tendency to revert to familiar, less inclusive processes, which clearly works against the long-term needs of ecosystem management. They argued that, as adaptive management demands a more open approach to uncertainty, and a more inclusive approach to decisionmaking, it creates a need for a person within the lead agency to act as a change agent: the institutional champion. Gilmour et al. (1999) claim that this person needs to be sufficiently influential in the lead agency decisionmaking processes to be able to keep the project focused on the experiment-review-feedback cycle essential to the implementation of an adaptive management strategy.

- Finally, Gilmour et al. (1999) noted that the development of a clear strategy of continuation of the adaptive management process is essential. The pressures against such a solution were high in these examples, and often will be because of financial, time, and human resource constraints.

In sum, the adaptive management applications discussed earlier found that the model development phase was useful in generating initial management proposals with substantial community interaction and support. However, it was the stage of embedding these models and approaches within existing management organizations (for use in longer term management processes) which appeared to be confronted by substantial barriers. Although community and stakeholder consultation, engagement, and interaction are important to an active adaptive management approach, it is indeed just one component. The key characteristic of an adaptive management approach remains embedding the capacity to test alternative response models over time, so that management organizations have ongoing objective information for making changes to policy settings. This crucial issue of embedding adaptive approaches in management organizations is returned to later in the chapter.

Current Applications of Adaptive Management Principles in Australian Water Management

The introduction to this chapter referred to the recognition of adaptive management within a wide range of Australian water management strategies, including within the National Water Initiative, and The Living Murray strategy. This section discusses the current and expected nature of the adaptive management process that will be employed in relation to these initiatives.

As has been outlined in earlier chapters of this book, separate intergovernmental agreements establishing the National Water Initiative (NWI) and progressing The Living Murray (TLM) were approved by the Council of Australian Governments (CoAG) in June 2004. The NWI intergovernmental agreement was signed by all Australian jurisdictions by April 2006. This agreement establishes a policy framework for ensuring greater compatibility and adoption of best management approaches to water management on a national scale (CoAG 2004), and is backed by an implementation budget of $2 billion. The second intergovernmental agreement is between the Australian government and the governments of New South Wales, Victoria, South Australia, and the Australian Capital Territory, and sets out arrangements for investing $500 million over five years to address water overallocation in the Murray-Darling Basin (CoAG 2004).

The Living Murray. The TLM Business Plan expressly requires an adaptive management approach. Although the development of the TLM strategy did not explicitly adopt an adaptive management methodology, there are a number of characteristics of the method that has been adopted which are consistent with an adaptive management philosophy. These include: a substantial process of technical investigations from a range of disciplinary perspectives (covering a range of physical and social sciences), in concert with ongoing community consultation processes with the wider basin community; the development of a range of options to be evaluated (including the 350GL, 750GL, and 1,500GL enhancement of environmental water allocations); the attempted integration of hydrological and ecological information through the Murray Flow Assessment Tool (MFAT)[3]; the provision of substantial financial and other resources to implement the preferred strategy as an experiment; and the commitment to establish the required monitoring and evaluation framework needed to support learning and adaptation during implementation of the strategy.

Criticisms of the TLM process from an adaptive management perspective would include the somewhat clouded political interactions leading to the selected strategy, (see Crase and O'Keefe, Chapter 11, and Rolfe, Chapter 13) and the passive adaptive management approach embodied by seriously considering only a single accepted response model as the basis of the evaluations. In May 2006, the Australian government announced that $500 million would be expended on improving the ecological health of the Murray-Darling Basin (an earlier commitment of $200 million was made in 2004). This will be used to complete capital works and undertake other measures to return 500 GL/year in additional environmental flows. Although this commitment is a positive step for the health of the Murray-Darling Basin, the size and nature of this intervention is not a consequence of any purposive evaluation of the impact of the initial strategy. This type of adaptive management is firmly of the evolutionary form.

The National Water Initiative. The NWI also explicitly requires an adaptive approach in the management of surface and groundwater systems as a key outcome of the initiative's water access and entitlements framework. In Chapter 4, Quiggin identified two major principles contained within the NWI that are important in dealing with risk and uncertainty in water management: that water access entitlements should be stated as shares of available water, rather than as specific volumes; and the specification of a risk assignment framework that allocates risk and uncertainty between governments and access entitlement holders according to whether water access changes are necessitated by natural events, by improvements in knowledge about water systems, or by changes in government policy.

The previously mentioned risk assignment framework, however, provides little real improvement in clarity about the allocation of risk and uncertainty, because of potential attribution problems. The legal perspective offered by McKay (Chapter 4) also supports this view. The NWI risk assignment framework may simply shift the focus to an unproductive debate about whether specific management changes are fundamentally driven by improvements in knowledge, natural climatic variability, or an underlying change in community objectives. In many cases the appropriate assignment of risk may not be able to be clearly determined. For example, if future research discovers that a particular organism is critical to riverine health, does the creation of standards requiring maintenance of this (hitherto unknown) organism at prescribed levels amount to a new environmental objective, or is it merely a change required by an improvement in knowledge? The answer is critically important, because it determines who bears the costs of change under the NWI risk assignment framework.

A related aspect of the NWI which has caused concern, because of possible adverse implications for an adaptive approach to water management, has been the strengthening of property rights by making water access entitlements perpetual in some circumstances. Concern about the strengthening of property rights emanates from the perception that, when property rights are stronger, there will be significant additional costs to government associated with the inevitable future changes that are made under an adaptive management regime (Productivity Commission 2003). Pagan and Crase (2005) assess potential interactions between the specification of water rights and the adaptive management philosophy, to see if property rights could be strengthened in ways that enhance adaptive capacity in water resource management. It was concluded that the strengthening of property rights over water could enhance adaptive management capacity in many circumstances, because while some costs may be increased, other static and dynamic costs of change can be substantially reduced.

There is a significant focus on monitoring and evaluation within the NWI, including the intention to develop an enduring national water information register—the Australian Water Resources Information System. However, at this stage there is no clear evidence of how a purposeful adaptive management philosophy will be incorporated into the implementation of the NWI. That is, there is an apparent commitment to provide comprehensive information systems, but a political rhetoric lingers which proposes that the NWI will fix Australian water management problems. Instead, an adaptive management philosophy requires a focus on developing a robust system which is able to more efficiently address ongoing problems as they become evident.

What Are the Opportunities and Barriers to the Future Incorporation of Adaptive Management Principles?

This section considers some of the key opportunities and barriers that will influence the capacity to improve the adaptive management of Australian water resources in the future.

Technological Advances Improving the Capacity to Provide Timely and Accurate Monitoring Data

There continue to be huge advances in technological capabilities to support monitoring and evaluation activities, as well as reductions in the cost of established technologies.

Technological advances in simulation modeling, remote sensing, software engineering, spatial analysis, automated instrumentation, and web information services have enabled proposals such as the Water Resources Observation

Network (WRON) to be developed. This initiative aims to provide web-based access to real-time data (using consistent standards) at a national scale. The WRON integrates censoring, data integration, automated forecasting, and information reporting aspects so as to provide timely information in a form that is useful to water resource managers (CSIRO 2006).

In addition to the monitoring of physical data (hydrology, water quality, etc.), the WRON proposal would also monitor socially relevant data such as changes in ownership of water access entitlements, where they are being transferred and so forth. This generates a more comprehensive base of information about policy effects (intended and unintended) which are associated more with social attitudes and human responses than with ecological linkages and responses.

These types of monitoring technology advances provide significant opportunities for the increased application of adaptive management, given the important nature of comprehensive and timely physical information to support learning as we go.

Institutional Barriers to Adaptation Within Water Management Agencies

The evidence of adaptation in Australian water management arrangements considered in this chapter has identified that, although NRM legislation and policies specifically require the use of adaptive management principles, it is rarely the case that their implementation has followed the more active forms.

Many other reviewers who have evaluated the success of adaptive management approaches have located the most significant barriers within the institutional architecture of resource management bureaucracies and institutions themselves. These authors also claim that the key challenge is to develop sufficient capacity within management institutions to be able to meet the requirements of an active adaptive management approach (Walters and Holling 1990; McLain and Lee 1996; Dovers and Lindenmayer 1997; Lee 1999; Johnson 1999a; Light 2002). The barriers to adaptive management within policy development institutions and managerial rewards systems that are relied upon throughout Australia for policy implementation include: problems in the acceptance of policies as experiments; problems in getting managers to respond to surprises and pursue the implications (rather than treat them as an aberration in the system); and problems in explaining/extending new information and gaining acceptance amongst stakeholders. Blann and Light (2000, *11*) conclude that "in a world dominated by novelty and change which we now inhabit, innovation, flexibility, mobility, and resilience are the attributes needed by our institutions to foster successful adaptation"—unfortunately, these features are not always valued within water bureaucracies.

Recognition of the Role of the Social Environment in the Management of Water Resources

A positive development that will enhance the potential for adaptive management of Australian water resources is the greater recognition of the role that social and economic circumstances play in the development of successful resource management strategies and management interventions. Important recent developments in this area in Australia include the instigation (in 1999) and expansion of the Social and Institutional Research Program (SIRP) by Land and Water Australia. The relevance of research in this area for adaptive management includes advances in the practical application of adaptive management (stakeholder consultation, organizational ownership and the like) as well as generating policies that embrace and reflect changing social objectives (in addition to changing biophysical circumstances). Notwithstanding these advances, the observations by Syme and Nancarrow in the following chapter show there is still room for improvement on this front.

The Net Benefits of Adaptive Management

Finally, a key attribute that will determine the extent to which more sophisticated forms of adaptive management (especially active adaptive management), are deployed in managing water resources in the future is the net benefits that can be generated from these methods relative to more conventional evolutionary adaptive management approaches.

Active adaptive management requires substantial investments of financial and other resources. This arises because of the innate requirements of the methodology in considering a range of response models, undertaking extensive monitoring and evaluation actions, so as to facilitate continuous improvement in the management outcomes. In some circumstances, the additional resource requirements of the approach will simply not be worth the potential improvement in resource management outcomes that may be possible.

In making these judgments, however, it is important to consider the full range of benefits and costs that may be relevant in the application of an adaptive management approach. These often include significant hidden benefits, including advantages in terms of both the opportunity costs and transaction costs of policy adaptation when inevitable changes in system dynamics occur.

Conclusion

The overall goal of adaptive management is *not* to maintain an optimal condition of the resource, but to develop an optimal management capacity. This is accomplished by maintaining ecological and social resilience that allows the system to react to inevitable stresses, and by generating flexibility in institutions and stakeholders that allows managers to react when conditions change. The result is that, rather than managing for a single, optimal state, we manage within a range of acceptable outcomes while avoiding catastrophes and irreversible negative effects (Johnson 1999b).

Implementing active adaptive management principles can improve the long-term efficiency of water resource use in comparison to reliance on weaker forms of adaptive management. An additional benefit of embracing an active form of adaptive management is that the information generated through experimentation and the testing of alternative hypotheses may allow probability distributions to be generated about the likelihood of a range of outcomes that were previously unknown. That is, some aspects of water management may be able to be treated as risk related issues (for which there are a broader range of mechanisms to manage, as outlined by Quiggin in Chapter 5), rather than being issues that are surrounded by uncertainty.

Active adaptive management concepts have been demonstrated to be particularly effective in generating new approaches to water management in difficult individual Australian situations; however, efforts to institutionalize those successful first stages have encountered significant barriers in instilling an ongoing active adaptive management philosophy within the relevant water policy development and implementation institutions.

Notes

1. These sections borrow heavily from the earlier work by Pagan and Crase (2005).

2. The NSW government's objectives also coincided with commitments made at a national level through the 1994 Council of Australian Governments (CoAG) strategic framework for the reform of the Australian water industry, and through the Murray-Darling Basin Ministerial Council regarding the Cap on Diversions detailed in earlier chapters.

3. The MFAT is a computer-based simulation model, and was used to integrate separate assessments across ten zones of the River Murray and its tributaries. MFAT was used to analyse alternative scenarios of how the environmental water might be used for each of the 350GL, 750GL, and 1,500GL levels of increased environmental allocation. These alternative scenarios included differences in flow conditions, accumulation and release strategies, as well as differences in the localities that the water was targeted toward. Outputs of the model include impacts on the condition of habitat for floodplain and wetland vegetation, waterbirds, and native fish, and on the growth of algal blooms.

References

Allan, C., and A. Curtis. 2003a. Learning to implement adaptive management. *Natural Resource Management* 6(1): 25–30.

Allan, C., and A. Curtis. 2003b. Notes from an adaptive management workshop, Lake Hume, July 2002, Johnstone Centre.

Barr, N., and J. Cary. 1992. *Greening a brown land: The Australian search for sustainable land use.* Melbourne, Macmillan Education Australia.

Blann, K., and S. Light. 2000. The path of last resort: Adaptive Environmental Assessment and Management (AEAM). Minneapolis: Adaptive Management Practitioners Network.

Council of Australian Governments. 2004. Intergovernmental Agreement on a National Water Initiative. www.coag.gov.au/meetings/250604/iga_national_water_initiative.pdf (accessed June 28, 2004).

CRC for Catchment Hydrology. 2005. Water quality models: Sediments and nutrients. *Series on Model Choice*, CRC for Catchment Hydrology.

CSIRO. 2006. Land and water: Environmental sensing, prediction and reporting. Canberra: CSIRO.

Dovers, S. R., and D. B. Lindenmayer. 1997. Managing the environment: Rhetoric, policy and reality. *Australian Journal of Public Administration* 56(2): 65–80.

Gimour, A., et al. 1999. Adaptive management of the water cycle on the urban fringe: Three Australian case studies. *Conservation Ecology* 3(1): 11.

Hax, A. C. and D. L. Wilde. 1999. The Delta model: Adaptive management for a changing world. *Sloan Management Review, Cambridge* 40(2): 11–29.

Holling, C. S., ed. 1978. Adaptive environmental assessment and management. International series on applied systems analysis. Chichester: John Wiley and Sons.

Johnson, B. L. 1999a. Adaptive management: Scientifically sound, socially challenged? *Conservation Ecology* 3(1): 10.

Johnson, B. L. 1999b. The role of adaptive management as an operational approach for resource management agencies. *Conservation Ecology* 3(2): 8.

Lee, K. N. 1999. Appraising adaptive management. *Conservation Ecology* 3(2): 3.

Lessard, G. 1998. An adaptive approach to planning and decision-making. *Landscape and Urban Planning* 40:81–87.

Light, S. 2002. Adaptive management: A valuable but neglected strategy. *Environment* 44(5): 42–43.

McLain, R. J., and R. G. Lee. 1996. Adaptive management: Promises and pitfalls. *Environmental Management* 20(4): 437–448.

MDBC (Murray-Darling Basin Commissin). 2005. Living Murray business plan. Canberra: Murray Darling Basin Commission.

NSW Water Management Act (2000).

Pagan, P., and L. Crase. 2005. Property rights effects on the adaptive management of Australian water. *Australasian Journal of Environmental Management* 12(2): 77–88.

Productivity Commission. 2003. *Water rights arrangements in Australia and overseas.* Melbourne: Productivity Commission.

Taylor, B., et al. 1997. Adaptive management of forests in British Columbia. Victoria, BC: Ministry of Forests.

Walters, C. J. 1986. *Adaptive management of renewable resources.* New York: McGraw Hill.

Walters, C. J., and C. S. Holling. 1990. Large-scale management experiments and learning by doing. *Ecology* 71(6): 2060–2068.

CHAPTER 15

The Social and Cultural Aspects of Sustainable Water Use

Geoffrey J. Syme and Blair E. Nancarrow

*I*T HAS LONG BEEN RECOGNIZED THAT the management of water resources needs to be planned with the social aspect of public good firmly in mind (Field et al. 1974). Indeed, there is now a substantial literature which attests to an abiding interest in the need to keep social factors in balance with environmental and economic considerations when thinking long term about our water resources. These interests have varied from discipline to discipline and from context to context, but their significance is nowhere more evident than when the mismanagement of water has been linked to the failure of past civilizations (Weiss et al. 1993) or identified as the genesis of water wars. Alternatively, the potential for enhanced cooperation between nations by establishing a better balance between social and other needs has been a theme of interest (see Biswas 1991; Caponera 1985 for differing viewpoints). Many see management of scarce water resources as a major factor influencing our health and mortality unless supply and demand is balanced (Stikker 1998).

These international perspectives have been reinforced in Australia. Despite the early warnings of some (Davidson 1969), prior to the last twenty years the mentality regarding water resources has been far from the issues associated with catastrophe, but more in line with the potential of water to boost regional development and the economy. Australia has a vast land mass and a relatively small population. The sky has been the limit and there was a perception that these few people could still make the deserts bloom. This part of Australia's water history cannot be ignored and is well summarized by Musgrave in Chapter 3. The development of water resources has been viewed in a positive political light for much of the twentieth century (Smith 1998, *190–191*).

The reality is, however, that while water development led to regional expansion toward the end of the century, it also became evident that Australia's water resources were finite and the environmental effects of overexploitation were increasingly obvious. Competition was emerging between user groups. For this reason, in 1992 the Council of Australian Governments (CoAG) commissioned a joint federal-state working group to report on measures to bring about an efficient and sustainable Australian water industry. Many aspects of this report were ultimately adopted by CoAG in 1994 (Smith 1998, *271*) and these are succinctly summarized by McKay in Chapter 4. Essentially, the thrust of the recommendations was to achieve triple bottom line sustainability (economic, environmental, and social) by using markets to move water to the use of highest economic value while protecting the environment by limiting human consumptive use. However, the third social metric was relatively de-emphasized in the early period of reform. The recommended changes to the economy and the environment were supposed to be delivered within social constraints. These constraints were never defined, although there was to be an emphasis on consultation and public education. Importantly, treating social norms as an input to water resource policy was, and continues to be, largely undeveloped.

Thus, initially social goals of water reform seemed to be restricted to the avoidance of unacceptably negative social impacts. It was not until recently that CoAG began to examine social issues in a systematic fashion. Given that reform had already begun to be implemented in the environmental and economic domains, it now seems vital that our understanding of social planning as it relates to water resources management is upgraded. Hopefully, this can occur before we commit ourselves, inadvertently, to a path that results in irreversible adverse social outcomes.

The recognition of the need for social planning in water resources management is not novel to Australia, nor is the problem of attaining it (Barriera 2003). To realize the extent of these problems one only needs to consider the Australian soldier settlement schemes after the First World War (see Musgrave Chapter 3). There has long been an understanding that water can be a direct determinant of many social benefits. Perhaps the thought of using water resources for social welfare has disappeared in preference to modern economic rationalism and the tendency for governments to refrain from intervening in society generally. Nevertheless, best practice in natural resource management now demands that social considerations have a systematic and high profile, both in terms of outcomes and the processes of decisionmaking.

This chapter explores the social dimensions of water policy in Australia and therefore offers a useful complement to the historical, legal, and economic perspectives which have dominated the earlier chapters. The role of community decisionmaking and the processes of problem solving should be considered in the context of the institutional and legislative frameworks outlined in earlier chapters. More specifically, the role of determining the appropriate water shares provides an important backdrop to the social analysis offered here.

Throughout the chapter we present a number of challenges derived from the literature and reflect on our own professional experience in this context. Comments on the weaknesses of social policy should not be construed as a criticism of what is happening now. Rather, these observations are proffered as a foundation for creating discussion and with the objective of improving future practice.

The chapter itself is arranged into seven additional sections. Section 1 provides a brief international context for the social elements of water policy formulation. Section 2 focuses on the benefits of water use as a vehicle for introducing the range of social components to water allocations. The following four sections are then devoted to specific challenges associated with the integration of social considerations into Australian water policy. The final section of the chapter considers mechanisms for developing a more inclusive framework in which social issues can be incorporated.

The International Context of Socioenvironmental Policy

In the context of the current discussion the Rio Declaration on Environment and Development by the United Nations (1992, Annex1, pp1, 3) specifies the following principles:

Principle 1

Human beings are the centre of concerns for sustainable development. They are entitled to a healthy and productive life in harmony with nature.

Principle 10

Environmental issues are best handled with participation of all concerned citizens, at the relevant level. At the national level, each individual shall have appropriate access to information concerning the environment that is held by public authorities, including information on hazardous material and activities in their communities, and the opportunity to participate in decisionmaking processes. States shall facilitate and encourage public awareness and participation by making information widely available. Effective access to judicial and administrative proceedings, including redress and remedy, shall be provided.

In Europe these overriding principles have been enshrined in the Aarhus Convention which emphasizes "the rights of access to information, public participation in decision-making, and access to justice in environmental matters" (UNECE 1998).

A community-based approach also satisfies the requirements of national best practice policies, released by Environment Australia (2002), which are similarly applied in the states. For example, in Western Australia, the Environmental Protection Authority (WA EPA 1999) and the Department of Environmental Protection (WA DEP 2003) both espouse the requirement for community involvement.

Despite these laudable aspirations for social inclusion and sustainability, the gap between rhetoric and practice remains large worldwide. Even in developed democratic countries the environmental justice movement has consistently demonstrated inequity in exposure to environmental problems. In terms of bearing the costs of environmental degradation and natural resource shortage, the poor and vulnerable have been consistently identified as being unfairly disadvantaged

(Faber 1998). This has sometimes led to the rejection of environmental groups by community interests representing the poor. In the United States at least, a highly effective coalition of social issue community groups has been formed which provides advice on a regular basis to the U.S. EPA on environmental justice (Faber 1998). No such organization exists in Australia, which is a pity, as allocation and sharing problems are becoming increasingly contentious.

In the case of widespread reallocation of water to the environment and the introduction of water markets, the social impacts of these measures have been scarcely considered, if at all in many places. For example, some research has shown that the introduction of water markets may have created poverty traps in some areas (Bjornlund and McKay 2000). Social analysis and discussion early in the planning and implementation process may have highlighted these potential effects.

Notwithstanding that there has been some limited recent attention focused on social issues, there is a need for a systematic focus on a number of social dimensions relating to both the culture of the decisionmaking community, and the rural and urban water users (in other words, the entire community). But as a first step it is important to be clear about what water means and what we are actually distributing.

There is a tendency for water allocation arguments to centre on the sharing of volumes of water. From a social perspective, it is important to recognize that we are, in fact, sharing the *benefits* that flow from access to water and its use (Moran et al. 2004). These do not always equate with volumes. The same body of water as it passes through a catchment can confer many benefits, and for this reason there are a myriad of ways in which water can be managed to gain a range of different social returns. Unless the scope of water accounting is sufficiently broad, some of these returns may be overlooked.

The Benefits from Water Use

Access to water affects the quality of life for people at all levels, from survival to the arcane levels of spiritual considerations. Jung's (1920, 1954) seminal psychological theories, for example, frequently allude to water imagery. Therefore, in social terms, the whole gamut of what it means to be human requires coverage by water planners.

It is evident that those needs that are most often investigated in quantitative (or any) terms are utilitarian either in economic or social contexts. Planners in Australia, brought up in a culture of economic rationalism, generally find the results quicker and easier to assimilate into traditional planning frameworks, such as cost-benefit analysis. This is possibly why economists have attempted so-called nonmarket valuations of some of the more humanitarian dimensions, as dollars are easier to incorporate in the decisionmaking process than, for example, scores on attitudinal scales.

Nevertheless, for most social scientists, the consideration of more complex water needs (e.g., spiritual, moral and cultural, or aesthetic needs) gives rise to demand for qualitative data, and often long term investigation (Strang 2004). The results of such studies are often seen to be in the academic or political sphere and too esoteric to incorporate in a rational planning process.

The tentativeness about social analysis is understandable. As with biophysical data, there are complex issues relating to levels of analysis and their interpretation that need to be worked through. In the social case, how does one scale up from local to regional to national levels? If we wish to create socially sustainable water resources management we also need to understand that time is a dimension (Ten-Houton 2005). We have to interpret this underlying benchmark against which social variables are being measured. Time and any associated changing benchmark can often be ignored in cross-sectional studies of attitudes.[1] This is to the detriment of understanding what the data mean in terms of social history and long term aspirations of the relevant community. For example, in a study in a Northern Tasmanian catchment (McCreddin et al. 1995), expectations about what the local rivulet should deliver depended on one's historical perspective. Younger members of the community were happy with the ability to fish for introduced species of trout. These were the good old days. By way of contrast, older members of the community could remember catching native black fish, and they wished to return to their different conceptualization of the good old days.

Although this appears simple and might be seen as a fundamental example of evolving perceptions, the immediate question is which baseline should be employed to adjudge aspirations of the community and why? To decide requires some technical investigation about whether black fish can be returned, but just as importantly we need to study how to combine the old and young perceptions and desires into acceptable and feasible long term goals for the community. This is why an understanding of social history is very important in the setting of long term goals, even if they are changed in the medium term. Clearly, these issues also determine what might be regarded as an appropriate share for the environment under the various state water planning schemes. A cross-sectional questionnaire is unlikely to capture the types of information required to provide an adequate answer to this complex question.

One of the major challenges for water resource managers over the next generation will be to find more acceptable ways to accommodate the evolving humanitarian needs within the planning processes. As highlighted by Crase and Dollery (Chapter 6) and Rolfe (Chapter 13), these changes give rise to transaction costs which can either promote or hinder policy reform. Anticipating and understanding these processes will not be easy, and will be beyond the current capabilities of most Australian agencies in the short term. It is perhaps because of the lack of suitably trained personnel in agencies that the social analysis of water resources management has often been partial, unsystematic, and generally lacking in theoretical rigor. It is firmly in the too-hard basket. If we consider future demands, it is even more vexing. However, the relatively small but growing literature from a variety of social science disciplines in Australia shows that, given the opportunities, social and cultural analyses can provide new insights and opportunities for getting the best out of Australia's water resources (see, for example, Jackson 2005). To do so requires a commitment to social inclusion in the water industry, and the willingness to confront a few basic problems in relation to Australia's water culture.

Here, we discuss four of these challenges, varying from the general to the specific. The first two concern the need for a social vision and the widespread reluctance to create tools for social interpretation which are relevant to the overall context for water management decisionmaking.

The latter two issues relate to implementation and are equally critical. Community concerns often relate to the fact that local knowledge is often not appreciated by decisionmakers and that risk is often not adequately considered by experts. Some of the cases examined by McKay (Chapter 4) provide a clear demonstration of this problem. Both of these concerns can lead to a lack of trust, which frequently provides the basis for sociopolitical confusion surrounding water resources decisionmaking. Risk and trust are critical to getting the decisionmaking process right from a social perspective. Implicit in dealing with these issues is the need to understand role differences between water professionals and the community, and their influence on the definition of the problem and assessment of preferred solutions.

Lack of Social Vision

Notwithstanding the fundamental significance of water to the welfare and functioning of urban and regional communities, there is very little planning to explore how water use and conservation can assist in creating desirable futures for Australian communities. Most national studies or audits content themselves with the collection of baseline information with some economic scenario projections.

Although, for example, France and other European nations acknowledge that the preservation of regional, agriculturally based communities is an inherent part of their culture and will act to protect them, there is a confused relationship between city and country in Australia. Some elements of this are evident in Musgrave's reflection on the history of irrigation development in Chapter 3.

Most of Australia's cultural icons relate to our relatively early history and to rural environments. For water, our images relate to flood and drought in the early settlement of the regions and our interior. Where water has been related to social engineering, it has been the development of irrigation systems and soldier settlements that have played this iconic role. The wisdom of soldier settlement has since been questioned, with the prospects for small landholders on the decline.

Today, the reality of Australian society is that it lives in cities and on coasts. Cities have the potential to compete with rural settlements and industries for a range of resources, including water. If this allocation issue is settled entirely by the market, urban communities have considerably more capacity to pay for the resource than rural communities. However, cities also have the capacity to create large environmental footprints if populations continue to rise and demand management and if alternative supply systems of water to metropolitan areas do not create the required efficiencies. Similarly, urban density and form are directly linked with quality of life. Green space and the role of water in the urban environment in generating well-being for urban residents are only now starting to be systematically considered by water managers and urban designers.

Although it is appreciated by many planners and policymakers that water resources underpin a wide array of human functioning—from protecting health, through economic development to support for recreation and finally spiritual values—there is no integrated framework for assessing the range of benefits from water in a planning context, nor for defining lexicographical choices (e.g., water allocation issues where trade-offs should not be considered for social or cultural reasons). Consequently, it can be argued that we have no water-driven social or cultural vision in Australia.

These problems are to some extent being tackled by governments. For example, there is a sustainability policy in Western Australia and a Sustainability Panel has been appointed to assess a proposal to transfer 40 gigaliters of water from the rural south west of Western Australia to offset the current shortage of water in the city of Perth. By and large, the approach of the Panel has been to follow the guidelines of Gibson (forthcoming). Whereas this is a worthy exercise, as social issues and impacts are being carefully scrutinized for the first time, it is difficult to interpret this social information in the absence of a clearly prescribed overarching social policy.

For example, it has been suggested by the Panel that although all water is a state resource and should be allocated in the best interests of the state, it should only be transferred if the reasonable needs of regional communities are protected. These reasonable needs are gradually being defined in terms of this specific proposal as the investigation continues. The understanding of how to frame the social investigation is also improving. But issues such as the long term integration of the Western Australian Water Supply system and the long term needs of the southwest communities remain problematic. Nevertheless, this learning-by-doing exercise is helping to develop a more holistic framework for incorporating social issues into decisionmaking and may be the precursor of improved social input into water reform in the longer term.

However, the problem remains that this project is being assessed in the absence of overarching social principles for the state. Western Australia shares this problem with most, if not all other states in the nation. The formulation and adoption of a State Water Strategy with community review in the near future provides the opportunity to fill this vacuum.

In summary, water resources management to date has been implemented without a social vision in Australia. However, recent recognition at federal and state levels that this is necessary for adequate planning means many governments and agencies will be faced with a steep learning curve.

The Prejudice Against Social Measurement

There is an ongoing perception among many decisionmakers that social science is subjective and that community preferences, feelings, and attitudes, even if measured quantitatively, are not sufficiently valid and reliable to be used to guide policy or planning. In any event, it is frequently believed that such measures are likely to change once the policy has been put in place and people recognize the wisdom of the planners (Dale et al. 1999).

In contrast, modeled economic analysis and forecasting based on industry economic statistics are most often seen as more legitimate routes to follow, as they are portrayed as nonsubjective—everyone knows the value of a dollar. However, such analyses also run the risk of limiting the range of benefits considered in benefit/cost judgments. For this reason economists and psychologists have led questionnaire-based research to expand economic analysis to nonmarket areas. Techniques such as contingent valuation and choice modeling have been developed as questionnaire techniques to elicit dollar values for variables such as environmental amenity in Australia and elsewhere (e.g., Bennett and Blamey 2001; Imber, Stevenson, and Wilks 1991). However, such techniques are often seen as being subject to bias, and their acceptance in policy circles has been highly variable.

Although there have been some commendable attempts to create triple bottom line indices for the water industry (Raskin et al. 1996), these have been only partially useful to decisionmakers, even when incorporated in systematic decisionmaking exercises. Frequently, the social indices are little more than extensions of economic indicators or merely demographic data (e.g., estimates of employment, or nonmarket values assigned to recreation). There are few attempts to incorporate social judgments of benefit outcomes to match the environmental and economic judgments.

This may be partly due to the difficulty of comparing combinations of indicators and attributes measured in different scales, across different beneficial outcomes and in different social groups. For example, how do you integrate attitudinal measures associated with water quality, and biodiversity indicators of water quality in any coherent and accountable way, especially if both indicators contribute to shared benefits as well as independent ones? More specifically, attitudes toward water quality and biodiversity indices may be closely correlated in pristine environments, but quite unrelated in modified urban environments. For this reason, we need to develop a technique that can evaluate and compare on the one scale the beneficial outcomes of water in each of the environmental, economic, and social domains for various policies.

One approach which offers some promise in this context is multicriteria analysis, and there are many outstanding examples of how this technique has greatly aided choice between differing water resource management options or scenarios (e.g., Pietersen 2006; Stewart and Scott 1995). But the important issue is gaining consistent levels of interpretation over differing case studies. A common lexicon of benefits rather than multiobjective ratings of individual benefits would assist.

Understanding the nature of a (water) benefit is a fundamental issue that has been grappled with over a number of years by a variety of disciplines, including economics and a wide variety of social science disciplines associated with social impact assessment (Baron 1998). No generally agreed consensus emerges from these disciplines on which variables are most important, or what set of indicators would be generally regarded as an adequate and comprehensive guide to benefits. For example, does the increase in the number of trees on all catchments lead to predictable benefits in human terms? If not, the decision to increase the number of trees needs to be assessed in terms of the benefits that accrue on a case-by-case basis. The use of human perceptions of benefits provides the common metric across catchments, and normative studies can be conducted using multicriteria or other valuation approaches.

Such a benefits metric would greatly aid our understanding as a nation of the overarching values which are guiding our decisions and clarify our vision as to what water should deliver for the Australian community. Work on developing such a benefit framework has begun and is in the early stages of implementation (Moran et al. 2004). Notwithstanding this progress, this initiative will not penetrate significantly into water resources management unless there is a greater involvement of social and integration scientists (for example, Jasanoff 1998).

Getting the Decisionmaking Process Right

Systematic and logical plans derived from professional knowledge and analysis (whether engineer's, scientist's, architect's, or economist's) are seen to be the way forward by those who think rational planning is best. Traditionally, many professionals don't see the community as rational. For these people, the public interest is thought to be protected by logical forecasting of demands which could be met by the best technical solutions. Nevertheless, it became evident by the 1960s that in order to define the public interest the community needed to be consulted (see, for example Newborough 1980).

Throughout the 1960s many countries witnessed the rise of the conservation and consumer movements. Against this backdrop, growth forecasts were questioned by communities who were to host certain developments. Protests became more overt and systematic. Thus, there was a rise in the phenomena of NIMBY (Not in My Back Yard) and LULU (Locally Unwanted Land Uses) protests in communities (Fort et al. 1993). At this point, largely because of frustration with apparently irrational community agitation, there was a rise in protest by the planners and developers themselves. They increasingly began to use the previously mentioned labels in their professional conferences and literature to stigmatize communities as irrational and selfish.

Many still consider that the public is basically ignorant and self centered and cannot be relied upon to make judgments for the public interest. In fact, the value of social data collection and public involvement is frequently questioned. An entrenched culture of self interest shared by all parties seems to have been developed (Miller 1999) and affects the ways in which social problems are addressed. This approach is broadly at odds with the growing and significant literature that shows that the public is capable of making judgments based on prosocial and justice criteria (e.g., Clayton 1996; Delli Priscoli 1998; Delli Priscoli and Llamas 2001; Kahneman et al. 1986; Syme 1998; Syme et al. 1999).

Arguably, there appears to be a mismatch between the espoused government commitment to public involvement and actual performance on the ground. This is demonstrated in the following figure from McCreddin et al. (1997) which shows that this shortfall is seen by both the public and organizational professionals themselves in at least one case study in the Australian milieu.

In Figure 15–1 the x-axis represents Arnstein's ladder of influence (Arnstein 1969). The bar chart shows clearly that, while the current level of public involvement was seen to be less than that desired, the preferred level of power was a partnership, rather than community domination of decisionmaking. It is interesting that the professionals also preferred such an arrangement.

The community in this study was also able to define what they meant as a partnership. In essence, this involved the setting of the planning criteria and the planner deciding on the best way to meet those criteria. The fear of some planners, that the public wish to replace them, is therefore unfounded in this example. It is partnership that is required and this has also been confirmed in other studies (see, for example, Nancarrow et al. 2002). The community is all too aware that they will not necessarily agree on all issues, that there will be conflict that requires careful management. There is also acknowledgment of the value of expert knowledge and the benefits of an external view of local issues.

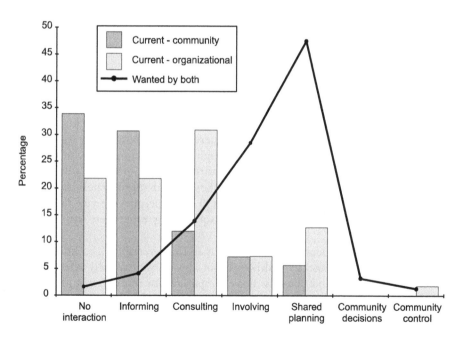

Figure 15–1. *Perceptions of Current Community Roles and Preferred Influence by Community and Professionals in Relation to Water Allocation in the Area*

Note: A single line is presented in terms of what is wanted by both parties because there was virtually no difference (McCreddin et al. 1997).

What is most interesting about this planning-oriented research is that the characteristics of partnership bear a remarkable resemblance to the social-psychology-based theory of procedural justice and the emerging area of fairness and its application to the natural resource management literature. The two approaches are beginning to merge at the community-based planning level.

Syme and Nancarrow (2002) developed a simple methodology that could evaluate the partnership or the extent of procedural justice in a wastewater planning process. In common with literature in other fields, it was found that the achievement of partnership led to greater commitment for future ongoing involvement with the agency. A burgeoning literature exists on the benefits of procedural justice and partnerships in decisionmaking processes. In sum, this literature points to a heightened degree of cooperation and the acceptance of decisions in natural resource management and other decisionmaking contexts (Tyler and Blader 2000). However, this same literature is largely either unknown or unused in water resources management. This is despite the similarity in findings with the experimentally based game theoretic literature (Wilke 1991). Given the lack of utilization of this community-based social justice research we are left with the common view of public involvement as inevitably increasing transaction costs and slowing decisionmaking. Yet we have some validated tools that can successfully include the public in decisionmaking, thus providing improved social process within the decisionmaking arena. The problem is that we have yet to systematically employ them.

Problem Solving: Dealing with Knowledge, Risk, and Trust

In this short section we deal with three key, largely unresolved issues within Australian water resources planning and management. These are knowledge, acceptable risk, and trust. Traditionally, there has been a tendency to imagine that if only the public were educated enough, took a rational view of risk and had the good sense to trust what the authorities were doing, the solutions to our water problems would be better, as they would appropriately reflect the expertise of water resources professionals. This is despite the fact that water always seems political and always worth squabbling about (see, for example, Davis 2001).

There is strong evidence in Australia that all three variables (i.e., knowledge, acceptable risk, and trust) can, although not always, substantially affect acceptance of water plans and projects. For example, recent research examining the tolerability of drinking water quality (Ross 2005) and willingness to tolerate wastewater reuse (Po et al. 2005) have shown both risk perception and trust in the water utility to be significant determinants of acceptance. Knowledge has, on occasions, also been a predictor of water-related attitudes, but the effects are variable. Finally, it must be noted that often risk and trust judgments are correlated and that they need to be managed in synchrony (see, for example, Das and Teng 2004). Managing these issues will be a key challenge for Australian water resource managers in the twenty-first century.

Lack of Understanding of the Role of Science Versus Lay Knowledge in Improving Our Water Resources Management

There is a preference among water resource professionals for systematic collection of scientifically defensible knowledge to assist with planning. Often locals feel that planners and scientists undervalue their lay knowledge and wisdom. However, knowledge needs to be interpreted within the planning context itself and the benefits that can potentially be delivered given the community's capacity and values. These factors are often defined in terms of the long term trends and/or indicators of significance to the community and/or a baseline case. Often much valuable information is held within a community to attest environmental change and the history of water management practices and community development. In this context it is worth noting the case of the *Murrumbidgee Groundwater Preservation Association v Minister for Natural Resources* detailed by McKay in Chapter 4. In this instance the Court settled in favor of the minister while also acknowledging that the groundwater sharing plan contained absurd, implausible, or irrational results from an irrigator's perspective. Conceivably, a less acrimonious process may have resulted from cooperative information sharing. One of the benefits of understanding community–government partnerships should be the better sharing of information to set strategic goals. Perhaps not surprisingly, quality information exchange has been shown to be a significant component of procedural justice (Lind and Tyler 1988).

Although local knowledge has recently become a key ingredient in conservation and planning in developing countries, it has not received the same degree of attention in Australia. Whereas there is growing interest in the incorporation of local input through companion modeling (Nancarrow 2005), and deliberative public processes (Proctor and Drechsler 2005), this is still not common. Moreover, it is clear that there is little theory about the role that local or lay knowledge can and should play in planning and policy, either in developing (Sletto 2005) or developed countries. This is despite the comments of Turnbull (1997, *551*) that:

> all knowledge traditions are spatial in that they link people, sites and skills. In order to ensure the continued existence of the diversity of knowledge traditions rather than have them absorbed . . . we need to enable disparate knowledge traditions in which the social organization of trust can be negotiated.

Knowledge management is a key area in need of development in Australia.

Agreeing on Risk and Uncertainty[2]

The lack of understanding of public involvement and confidence in social analysis is exacerbated when issues of risk or uncertainty are concerned. This is because the community and decisionmakers sometimes have different ways of constructing estimates of the risk associated with the environmental issue and the quality of decisions associated with it. Rather than improve our understanding of why these decisions are made differently, much of social science itself has concentrated on controlling outrage and explaining why people over- or underestimate risks with a view toward incorporating rationality into the debate (Baron 1998). There is, unfortunately, a much smaller literature explaining subjective biases in risk analysis among scientists themselves.

Renn (2004) notes that members of the public tend to concern themselves with the long term effects of risks, personal control, equity and fairness issues, and the rapidness of change. In contrast, the professional concentrates on minimizing the risk of adverse effects (risk meaning probability multiplied by consequence). These general observations have been found to be applicable in Australia for water resources issues, for example dam safety (Syme 1995). There are now numerous tomes on understanding and integrating the various perceptions of risk within the planning process. These treatises vary from the theoretical to the pragmatic (Morgan et al. 2002; Jaeger et al. 2002). Despite these resources and the ubiquitous occurrence of risk and uncertainty in decisionmaking, they are not regularly used as a basis for public input to decisionmaking. This is concerning because of their relationship to trust (see Bloomquist 1997 for the many definitions of trust). Sharlin (1989) suggests, for example, that the public meaning and acceptability of risk is actively changing because of the decline in trust in public agencies and of course vice versa.

The Growing Lack of Community Trust

There are reports in Australia and elsewhere of declining levels of deference or trust in government and the scientific community. This is possibly exacerbated by the lack of procedural justice in decisionmaking, and the perceived self interest culture that pervades policy discussion. Belief that another will place the public interest above their personal gain is less common (Miller 1999). The result is a decisionmaking system focused on catering for stakeholder groups' interests and competition rather than social vision and cooperation. The problem then becomes one of a system based on sectoral inputs which can collapse when there is no perceived individual incentive for participation. Some call this a social trap (Platt 1973); we are selfish because we are taught to be selfish.

In fact, there is substantial evidence that the public in Australia use the three dimensions identified by Wilke (1991) in his review of game theory research when considering water resources allocation issues; efficiency (water should not be given to those who waste it), greed (water should provide economic returns or perceived personal gain for the individual), and fairness (there should be both procedural and distributive justice in the allocation of water resources). Notwithstanding the arguably inflammatory nature of this typology, currently, by far the greatest emphasis is given to the notions of efficiency and greed in public decisionmaking processes. Fairness is largely ignored despite the development in Australia of techniques to include it in a transparent fashion (Syme et al. 1999). Failure to address issues of fairness can be a major impediment to change in both social and economic terms. For example, as discussed elsewhere in this volume, exit fees associated with permanent sales of water entitlements are currently being imposed in some irrigation districts to ensure that the risk of stranded assets is minimized. However, as noted by Crase and Dollery (2007), these arrangements may ultimately benefit better-off farmers and are unlikely to be economically efficient. This effect may be compounded by the social stigmatization of people who opt to sell water (Fenton 2006). As a result of this lack of attention to fairness, prosocially oriented concepts such as trust become eroded.

So What Should We Be Doing to Give Social Issues a Place at the Table?

It is clear from the policies both nationally and internationally that there have been laudable attempts to provide a social underpinning to sustainability in water resources management. Nevertheless, it is also apparent that, despite best intentions, there is a gap between promise and implementation. This is not to undervalue significant social contributions in some cases.

It is obvious that the social approach to water resources management in Australia is in urgent need of development. Fortunately, however, over recent years socially based tools have become available to assist in producing socially responsive decisionmaking processes and analysis. The big issue we face is in developing a community-sensitive culture to our water resources planning and management. To do this we need to raise questions that may cause some debate and conflict amongst water resources managers themselves.

First, we need to start a national debate on what water culture we wish to create and what are the important societal values we would wish water resources management to preserve. As part of that we need a greater understanding of the range and nature of beneficial outcomes from water and their relationship to predominantly economic evaluations of water policy issues. In fact, the relationship between social and economic analysis and decisionmaking processes deserves some organized debate. The value systems of both need to be recognized and the potential for synthesis examined (Zajac 1995).

Perhaps the simplest but most challenging task is to create a commitment to the development, implementation and evaluation of procedurally just decisionmaking processes. These processes will need to include choice techniques which are usable by both the community and the planner (Nancarrow and Syme 2001) and avoid the current common perception that the community is the guinea pig for black box methodologies that may or may not provide the answer that meets key social criteria. The black box perception is promulgated by the inclusion of algorithms or complex statistics in a model in a way which is not easily understood, rather than the use of interactive techniques to develop the model in partnership with the community (Nancarrow 2005). This is not rocket science but requires both commitment and ingenuity on behalf of both the planner and the community. Trust and partnership are required.

It must be noted that as Australians have understood more about their water resources, environmental sustainability alone has required reallocation of water from traditional consumptive uses to nonconsumptive uses and between individual enterprises within established sectors. Judgments about distributive justice attend reallocation decisions. Australian water policy is punctuated by calls for the resource to be moved to higher-value uses. What is not at all clear is whose values are to be used as the metric. The criteria against which these judgments are made need to be explicit and within a legible, coherent and procedurally just decisionmaking process. We need to know what and whose values are being represented in the discussions, who are the winners and losers, what is the perceived nature and implementation of the public good, and so on. To date we have been much less than comprehensive in our approach to sharing pain as well as gain. We have also only used two legs of Wlike's (1991) stool: greed and efficiency. Arguably, there is a case for greater consideration of fairness to add stability to the debate.

Finally, and most importantly, it is clear that to approach the general solutions discussed earlier will require dedicated resources both in federal and state government agencies. There will need to be a change in the skills base and even in the structure in some agencies, and the routine voice of both social scientists and the community at meaningful decisionmaking levels. This would bring the issues of knowledge, trust, and risk into clearer focus.

In the past this has not been easy. For example, part of the problem facing many agencies reportedly has been access to appropriate skills. Jobs in government social impact groups, for example, are sometimes difficult to fill. The increased training opportunities available in Australian universities will, however, in the very near future begin to resolve this issue. Our only problem now is to grasp the vision that it is better to plan for a positive social future than implement water policies and then attempt to ameliorate and describe the ensuing negative social impacts.

Notes

1. See Chapter 6 for a discussion of the transaction costs involved with dynamic assessment of evolving institutions.

2. Earlier, Quiggin (Chapter 5) analysed the economic dimensions of risk and uncertainty. Here we confine ourselves to the social conceptualization of these issues.

References

Arnstein, S. A. 1969. A ladder of citizen participation. *American Institute of Planning Journal* 32:216–226.

Baron, J. 1998. *Judgment misguided: Intuition and error in public decision making.* New York: Oxford University Press.

Barreira, A. 2003. The participatory regime of water governance in the Iberian Peninsula. *Water International* 28:350–357.

Bennett, J., and R. Blamey. 2001. *The choice modeling approach to environmental evaluation.* Cheltenham, UK: Elgar.

Biswas, A. K. 1991. Water for sustainable development in the 21st century: A global perspective. *Water International* 16:219–224.

Bjornlund, H., and J. M. McKay. 2000. Do water markets promote a socially equitable reallocation of water in Victoria Australia? *River* 7:141–154.

Bloomquist, K. 1997. The many faces of trust. *Scandinavian Journal of Management* 13:271–285.

Caponera, D. A. 1985. Patterns of cooperation in international water law: Principles and institutions. *Natural Resources Journal* 25:563–588.

Clayton, S. 1996. What is valued in resolving environmental dilemmas: Individual and contextual influences. *Social Justice Research* 9:171–184.

Crase, L., and Dollery, B. 2007. Exit fees and the myth of externalities in the transfer of irrigation rights. Paper presented at the Australian Agricultural and Resource Economics Conference, December, Queenstown.

Dale, A., N. Taylor, and M. Lane. 1999. *Social assessment in natural resource management institutions.* Melbourne: CSIRO Publishing.

Das, T. K., and B-S. Teng. 2004. The risk based view of trust: A conceptual framework. *Journal of Business and Psychology* 19:85–116.

Davidson, B. R. 1969. *Australia wet or dry: The physical and economic limits to the expansion of irrigation.* Melbourne: Melbourne University Press.

Davis, S. K. 2001. The politics of water scarcity in the western states. *The Social Science Journal* 38:527–542.

Delli Priscoli, J. 1998. Water and civilisation: Using history to reframe water policy debates and build new ecological realism. *Water Policy* 1:623–636.

Delli Priscoli, J., and M. R. Llamas. 2001. International perspective on ethical dimensions in the water industry. In *Navigating rough waters: ethical issues in the water industry,* edited by C. K. Davis and R. E. McGinn. Denver: American Water Works Association.

Environment Australia. August 2002. Best Practice Environmental Management in Mining. Commonwealth of Australia. http://www.ea.gov.au/industry/sustainabile/mining/booklets/overview/ch5.htm;#p5 (accessed November 22, 2005).

Faber, D. 1998. *The struggle for ecological democracy: Environmental justice movements in the United States.* New York: Guilford.

Fenton, M. 2006. *The Social Implications of Permanent Water Trading in the Loddon-Campaspe Irrigation Region of Northern Victoria.* Report prepared for the North Central Catchment Management Authority.

Field, D. R., J. C. Barron, and B. F. Long. 1974. *Water and community development.* Ann Arbor, MI: Ann Arbor Science Publishers.

Fort, R., R. Rosenman, and W. Budd. 1993. Perception costs and NIMBY. *Journal of Environmental Management* 38:185–200.

Gibson, R. B. Forthcoming. *Sustainability assessment: Criteria and processes.* London: Earthscan.

Imber, D., G. Stevenson, and L. Wilks, L. 1991. *A contingent valuation survey of the Kakadu conservation zone.* Canberra: Australian Government Publishing Service.

Jackson, S. 2005. Indigenous values and water resources management: A case study from the Northern Territory. *Australasian Journal of Environmental Management* 12:136–146.

Jaeger, C., O. Renn, E. Rosa, and T. Webler. 2002. *Risk and rational action.* London: Earthscan.

Jasanoff, S. 1998. The political science of risk assessment. *Reliability, Engineering and System Safety* 59:91–99.

Jung, C. G. 1920. *Collected papers on analytical psychology.* 2nd ed, London: Balliere, Tindall, and Cox.

Jung, C. G. 1954. *The development of personality,* trans. R. F. C. Hull. London: Routledge.

Kahneman, D., Knetsch, J. L., and Thaler, R. 1986. Fairness as a constraint on profit seeking entitlements in the market. *American Economic Review* 76:728–741.

Lind, E. A., and T. R. Tyler. 1988. *The social psychology of procedural justice.* New York: Plenum.

McCreddin, J. A., B. E. Nancarrow, and G. J. Syme. 1995. *Community Definition & Management of Environmental Flows: Clayton's Rivulet Case Study.* A report to the Department of Primary Industries and Fisheries, Tasmania. CSIRO Division of Water Resources Consultancy Report No. 95–28.

McCreddin, J. A., G. J. Syme, B. E. Nancarrow, and D. R. George. 1997. *Developing Fair and Equitable Land and Water Allocation in Near Urban Locations: Principles, Processes and Decision Making*. CSIRO Division of Water Resources Consultancy Report No. 96–60. Perth, Australia: CSIRO.

Miller, D. 1999. The norm of self interest. *American Psychologist* 54: 1053–1060.

Moran, C. J., G. Syme, S. Hatfield-Dodds, N. Porter, N. Kington, and L. Bates. 2004. On Defining and Measuring the Benefits from Water. Paper presented at 2nd IWA Leading Edge Conference on Sustainability—Sustainability in Water Limited Environment. November 2004, Sydney.

Morgan, M. G., B. Fischhoff, A. Bostrom, and C. J. Atman. 2002. *Risk communication: A mental models approach*. Cambridge: Cambridge University Press.

Nancarrow, B. E. 2005. When the Modeller Meets the Social Scientist or Vice Versa. Paper presented at MODSIM05 International Congress on Modelling and Simulation. December 2005, Melbourne, Australia.

Nancarrow, B. E., C. S. Johnston, and G. J. Syme. 2002. *Community Perceptions of Roles, Responsibilities & Funding for Natural Resources Management in the Moore Catchment*. CSIRO Land and Water Consultancy Report, Perth, Australia.

Nancarrow, B. E., and G. J. Syme. 2001. Challenges in implementing justice research in the allocation of natural resources. *Social Justice Research* 14:441–452.

Newborough, J. R. 1980. Community psychology and the public interest. *American Journal of Community Psychology* 8:1–17.

Pietersen, K. 2006. Multiple Criteria Decision Analysis (MCDA): A toll to support sustainable management of groundwater resources in South Africa. *Water SA* 32:119–128.

Platt, J. 1973. Social traps. *American Psychologist* 28:641–651.

Po, M., B. E Nancarrow, Z. Leviston, N. B. Porter, G. J. Syme, and J. D. Kaercher. 2005. *Predicting community behaviour in relation to wastewater reuse: What drives decisions to accept or reject?* Perth: CSIRO Water for a Healthy Country.

Proctor, W., and M. Drechsler. 2005. Deliberative Multi-criteria Evaluation. *Environment and Planning C: Government and Policy—Special Edition in Participatory Approaches to Water Basin Management*, forthcoming.

Raskin, P. D., E. Hansen, and R. M. Margolis. 1996. Water and sustainability: Global patterns and long range problems. *Natural Resources Forum* 20:1–15.

Renn, O. 2004. Perception of risks. *Toxicological Letters* 149:405–413.

Ross, V. 2005. *The Determinants of Trust and Satisfaction with Drinking Water Quality*. Perth: Australian Research Centre for Water in Society Occasional Paper.

Sharlin, H. I. 1989. Risk perception: Changing the terms of debate. *Journal of Hazardous Materials* 21:261–272.

Sletto, B. I. 2005. A swamp and its subjects: Conservation politics, surveillance and resistance in Trinidad in the West Indies. *Geoforum* 36:77–93.

Smith, D. I. 1998. *Water in Australia: Resources and management*. Melbourne: Oxford University Press.

Stewart, T. J., and L. Scott. 1995. A scenario-based framework for multicriteria decision analysis in water resources planning. *Water Resources Research* 31: 2835–2843.

Stikker, A. 1998. Water today and tomorrow: Prospects for overcoming scarcity. *Futures* 30:43–62.

Strang, V. 2004. *The meaning of water*. Oxford: Berg.

Syme, G. J. 1995. Community acceptance of risk: Trust, liability and consent. In *Acceptable risks for major infrastructure*, eds. P. Heinrichs and R. Fell, 31–40. Rotterdam, Netherlands: A.A. Balkema.

Syme, G. J. 1998. If water means wealth—how should we share it? In *Proceedings of the Irrigation Association of Australia National Conference and Exhibition*, 23–32.

Syme, G. J., and B. E. Nancarrow. 2002. Evaluation of public involvement programs: Measuring justice and process criteria. *Water* 29(4): 18–24.

Syme, G. J., B. E. Nancarrow, and J. A. McCreddin. 1999. Defining the components of fairness in the allocation of water to environmental and human uses. *Journal of Environmental Management* 57: 51–770.

TenHouton, W. D. 2005. *Time and society*. Albany, NY: State University of New York Press.

Turnbull, D. 1997. Reframing science and other local knowledge traditions. *Futures* 29:551–562.

Tyler, T. R., and S. L. Blader. 2000. *Cooperation in groups. Procedural justice, social identity and behavioral engagement*. Philadelphia: Psychology Press.

UNECE (United Nations Economic Commission for Europe). 1998. The Aarhus Convention http://europa.eu.int/comm/environment/aarhus (accessed,21st July, 2005)

United Nations. 1992. Report of the United Nations conference on environment and development Annex 1, pp1,3 http://www.un.org/documents/ga/conf151/15126–1annex1.htm (accessed June 20, 2005).

WA DEP. (Western Australia. Department of Environmental Protection). 2003. Air Quality in Western Australia. Ambient Air Quality Guidelines http://aqmpweb.environ.wa.gov.au:8000/air_quality/Policy_Legislation_and_Regulations/Ambient_Air_Quality_Guidelines (accessed July 30, 2003).

WA EPA. (Western Australia. Environmental Protection Authority). 1999. Developing a Statewide Air Quality Environmental Protection Policy. Concept Discussion Paper for Public Comment. http://epa/wa/gov/au/template.asp?ID=20&area=Policies&Cat=Environmental+Protection+Policies+%28EPP%29 (accessed July 30, 2003).

Weiss, H., M. A. Courty, W. Wetterstrom, F. Guichard, L Senior, R. Meadow, and A. Curnow. 1993. The genesis and collapse of third millennium Mesopotamian civilisation. *Science* 261:995–1088.

Wilke, H. A. M. 1991. Greed, efficiency and fairness in resource management situations. *European Review of Social Psychology* 2:165–187.

Zajac, E. E. 1995. *Political economy of fairness*. Cambridge, MA: MIT Press.

CHAPTER 16

Lessons From Australian Water Reform

Lin Crase

THE CHALLENGES THAT ARISE FROM formulating contemporary water policy in Australia touch all sectors of the community. Accordingly, an understanding of the multifarious aspects of water allocation and decisionmaking is required to advance our appreciation of this complex field. It is the aim of this book to provide some progress on this front.

However, as was acknowledged in the opening chapter, bringing together these disparate and yet related strands of thinking raises its own set of problems. In addressing these, the components of this book have been shaped around the following questions:

1. What is the magnitude and contour of water challenges in Australia?
2. What are the motivations that are driving water reform in Australia?
3. What is the ideal state of water resource management to which Australians aspire?
4. What are the key characteristics of reform, and the impediments to and consequences of the reform journey?
5. What lessons might be derived from the reform experience?

These questions are now revisited in an effort to marshal key findings and review enduring themes for future policy development.

What Is the Magnitude and Contour of Water Challenges in Australia?

We commenced our investigation of Australian water policy by examining the hydrological characteristics of Australia's water resources (see Letcher and Powell, Chapter 2). Core to the human reaction to Australian water resources has been the inherent variability of the resource, when measured in both quantity and quality terms. This variability occurs in two ways—spatially, which has given rise to marked disparities in use, infrastructure, and institutional responses within and across jurisdictions, and temporally, resulting in high intra-year and inter-year fluctuations. The latter form of variability has created unique challenges to managing the security of water supply in the Australian context. More specifically, as noted by Warren Musgrave (Chapter 3), Australian storages need to be significantly larger than elsewhere in the world to achieve the same level of supply security. This is accompanied by significant economic challenges for Australia's largest extractive water user—the irrigation sector. Arguably, until recently many of these challenges have been socialized in the form of policy that supported the broad concept of yeomanry and an environmental ethos that permitted the costs attending resource degradation to be carried by the wider community.

Notwithstanding the monumental hydrological hurdles to achieving resource development and security, Australia's water history since European settlement is punctuated by bold, publicly funded projects aimed at harnessing water for productive pursuits, often under the guise of policies like regional development or soldier settlement. Warren Musgrave (Chapter 3) has categorized these events into separate phases of water history as applied in the Murray-Darling Basin. More specifically, the late nineteenth and early twentieth centuries laid the foundation for the popularity of resource development as a policy approach. The riparian doctrine was abandoned; states assumed control of water resources; and the ground for establishing powerful water bureaucracies was prepared. Thereafter and until the 1960s (the second phase), water resources were enthusiastically tamed with expansive dam building and the creation of a significant public irrigation sector, particularly in the eastern states.

However, this broad approach to water resources was not limited to the Murray-Darling Basin. In Tasmania, a similar approach manifested in the expansion of the hydroelectricity sector (see Duncan and Kellow, Chapter 8). In other states the rampant mining of the Great Artesian Basin subsequently resulted in a series of untapped bores running freely as if the resource were infinite (see McKay, Chapter 4). In Western Australia, irrigation development was less prominent but has nevertheless attracted scrutiny during the third reformist phase of water history which stands incomplete.

This most recent phase has grown from public discontent with the economic performance of the various extractive water sectors and the recognition that environmental harm is inextricably linked to our efforts to secure a water resource which is inherently variable and uncertain under natural conditions. Fueled by a wider trend toward more prudent use of the public purse and mounting concerns about the prospects of anthropogenic climate change, the challenge for policymakers has been to transform institutions, both formal and informal, into those that accord with an acceptable water management ethos. This is a formidable task in its own right, even if the policymaker was commencing with a blank sheet. In reality, the challenge is aggravated by path dependencies, the legacies of earlier resource allocation decisions, inter-jurisdictional rivalries, and incomplete knowledge. Put simply, the challenge of water policy in Australia is not one of variable quantity and quality of water per se—it is a social problem requiring adjustments in the behavior of competing groups, the systematic accumulation of knowledge, and the enactment of adaptable solutions that can be amended as increased understanding reveals new challenges.

In this context, examination of Australian water policy provides a lens into other policy arenas. It also provides insights into Australian society generally as governments struggle to find resolutions to resource reallocation problems when an established set of property rights has been put in place.

What Are the Motivations That Are Driving Water Reform in Australia?

Two sources have been consistently cited as driving water reform in Australia. The first of these relates to the enhanced standing of the natural environment in public policy. Jennifer McKay (Chapter 4) briefly detailed the principles that pertain to Ecologically Sustainable Development (ESD) that were ultimately enshrined in legislation in all states and territories. These principles have also buttressed the national trend toward ascribing adequate water to achieve environmental ends. In addition to the legal complexities and inconsistencies attending this approach, there are critical deficiencies in the social interpretation of environmental values, some of which have been identified by Syme and Nancarrow (Chapter 15).

The second driver of reform has emanated from concerns about the economic treatment of water resources and the performance of extractive users when faced with subsidized supply, particularly in irrigation. Moreover, these concerns have been characterized as indicative of a broader acceptance of the merits of reduced intervention by government along with expanded approval of the paramountcy of markets and individualism (see Crase and Dollery, Chapter 6). In this context the motivations for reform represent a two-dimensional movement; one indicated by the transition from developmentalism to sustainability and the other from state control of resources to individualism.

In both cases these changes symbolize a response to the discord between social norms and the formal policies of government. In this context, the motivations for reform have been cast in terms of transaction costs (see Crase and Dollery, Chapter 6). Put simply, water policy reform is occurring because the costs of the status quo (i.e., the mismatch between the efficient allocation of the resource—in both productive and allocative terms—and the current allocation) are considered intolerably high against the cost of policy change.

There are two potential procedural aspects to these changes. First, the bureaucracy could be conceptualized as an omniscient and benevolent entity guiding and steering the community to a preferred state. Second, the community itself can be pictured as agitating for change via the political process, thereby forcing a recalcitrant water bureaucracy to accept the inimitability of environmental and economic forces. Whereas the reality is probably a combination of the two, there is a marked difference in the part played by the bureaucracy in the various water use sectors—the bureaucracy still holds considerable sway in the urban sector, for example, as noted by Edwards (Chapter 10).

Importantly, there is no convincing evidence that equity has reached the status of a policy goal in its own right within the Australian water debate. Concerns about fairness in resource allocation are considered by Geoff Syme and Blair Nancarrow in Chapter 15; however, finding the appropriate policy response to this issue is no simple task. Equity is not easily defined, particularly in the context of water. Poignant questions in this context include: Is it equitable to require one sector (irrigation) to significantly reduce its consumptive use of water after decades of public policy which encouraged that sector's expansion? Is it fair to (*prima facie*) enhance the individual rights of irrigators in order to establish a market framework? Coase has long argued that the initial distribution of rights is irrelevant because, once established, the market will give rise to efficiency-enhancing transactions. With that said, the political dimension to this problem is a critical determinant of the wider acceptability of this approach. Related but equally challenging are questions like: Do those Australians who hold strong concerns about the degraded state of the nation's rivers consider the slow pace of reform to be fair? Would urban Australians continue to broadly support punitive water restrictions if they realized there were alternative and more efficient approaches to water rationing? Many of these policy questions are raised in earlier chapters. Readers attempting to answer these questions might not be surprised to observe that, to date, Australian policymakers have primarily concerned themselves with the notion of procedural fairness (see Crase and O'Keefe, Chapter 11) or have sought to build the perception that fairness has been afforded to the policy formulation process. Similarly, McKay's (Chapter 4) observation that most of the cases dealing with the water planning process are likely to be tested under administrative law, and thereby avoid the thorny issue of administrative justice or error, further demonstrates the difficulties of addressing this issue.

What Is the Ideal State of Water Resource Management to Which Australians Aspire?

There are two major directions embedded in contemporary water policy in Australia; that is, the simultaneous movement toward sustainability in environmental terms and the drive toward efficiency within the consumptive sectors by promoting individual decisionmaking. Arguably, a third element, equity, could be included on another plane (as has already been noted), but the extent to which this could be adequately defined, given our limited knowledge of preferences for equity, would render this largely meaningless.

There seems little doubt that Australians have come to expect more from public policy on the environmental front. In the context of water, this expectation has gone beyond simply limiting future developments that might potentially degrade natural ecosystems. Rather, there is now considerable momentum to restore degraded systems which, by definition, requires a reallocation of the resource away from some current consumptive users.

Similarly, the Australian enthusiasm for smaller government and market determination of resource allocation, which has been evident in several domains for the last quarter of a century, shows no signs of abating. State and national governments from both sides of politics continue to espouse their economic credentials in the form of larger fiscal surpluses and the expanded use of private sector decisionmaking in preference to state intervention.[1]

In November 2006, when inflows into the Murray-Darling Basin were the lowest on record and large tracts of inland and southern Australia were facing the worst drought in history, a water summit was hosted in Canberra. The major announcement that followed was not one of increased control by the state to address the obvious and growing water shortage. Rather, the prime minister announced that interstate permanent water trading between states in the southern Murray-Darling Basin, which was initially accepted as part of the National Water Initiative, would be accelerated to see the realization of a national water market by January 2007. Such policy faith in the capacity of the market to deal with resource scarcity is arguably indicative of the wider community acceptance of the role of market forces.

However, faith in the market is not enough. As several authors have noted, the specification of property rights to water is incomplete in Australia and activation of water markets continues to reveal deficiencies. Quiggin (Chapter 5) observed the early impact of sleeper and dozer rights on the security of supply. The rights to return flows from irrigation are also not specified and their prominence in underpinning flows is not well understood (Crase and Dollery, Chapter 6). Urban users' rights to water are attenuated by urban water authorities but there are serious questions about the knowledge base upon which the structure of those rights has been based (Edwards Chapter 10). Individual irrigator's rights are being amended in reaction to the perceived and potential impact on community-funded irrigation infrastructure (McLeod and Warne, Chapter 7). The rights to an acceptable level of amenity in the receiving environment for treated urban waste water is not well defined, thereby confusing the price signals to encourage urban reuse (Khan, Chapter 12). Collectively, these deficiencies leave policymakers often playing catch-up by introducing administrative caveats and constraints on-the-run. The decision by the New South Wales government to suspend water trade in the Murray and Murrumbidgee Valleys early in November 2006 (partly because of concerns about the impact of trade on supply reliability) and then to effectively reverse that decision by agreeing to the accelerated development of the national water market in January 2007 is demonstrative of the problem. Moreover, accelerating the development of the national water market seems likely to stretch the capacity of legislators to keep up with the amendments necessary to head off perverse outcomes. This type of behavior falls well short of the systematic and robust adaptive management approach described by Pagan (Chapter 14).

At a broader level the trends in water policy potentially contain significant paradoxes and contradictions. Moreover, resolution of these problems is hampered by the absence of an established social metric capable of accounting for the inevitable trade-offs between competing goals (see Syme and Nancarrow, Chapter 15). Establishing a baseline of acceptable environmental objectives is a subjective process, regardless of the proclivity of some disciplines to portray it differently. Accordingly, identifying the desired location for the Australian community on the continuum between state control and individual control on the one hand, and resource development versus resource conservation on the other, is problematic.

Although it is conceptually convenient to consider water policy as a progression along two broad planes, this runs the risk of ignoring the importance of uncertainty that circumscribes water policy formulation. John Quiggin (Chapter 5) identifies the futility and folly of attempting to invoke policies that espouse the complete elimination of uncertainty. In the context of Australian water, it is important to realize that resource security is a rare commodity and is also ". . . in large measure, a zero-sum commodity. The more security is given to one group of users, the less there is for everybody else" (Quiggin, Chapter 5). The assignment of well-defined property rights offers some promise for overcoming this problem, but a comprehensive definition of rights also requires a more complete understanding of the resource than is presently available.

Extending this line of thinking, any attempt to articulate the aspirational location of Australians with respect to water also runs the risk of assuming that these values are stable and our knowledge of the measures required to achieve them is entire. Neither of these situations exists and the need for policy adaptability must be included as a broad goal. Unfortunately, a substantial divide remains between the rhetoric of adaptive management and its application: the use of a truly robust adaptive management approach is not yet evident in most water policy in Australia (see Pagan, Chapter 14).

Notwithstanding its egalitarian traditions, Australia is not a homogenous society and the acceptance of the various strands of water reform varies between communities. This is highlighted by the contrasting reactions in the various water use sectors (Chapters 7–10) and the potential for a growing divide between the opinions of rural and urban Australians (see Crase and O'Keefe, Chapter 12). Arguably, these phenomena and the lack of knowledge about Australia's water resources, at both the community and policymaker level, have resulted in some nontrivial policy inconsistencies (see Edwards, Chapter 10).

What Are the Key Characteristics of Reform, and the Impediments to and Consequences of the Reform Journey?

As we have noted, there are two separate and distinguishable policy motivations. First, there has been a deliberate effort to attend to the environmental stress emanating from fully allocated or overallocated systems. Even in Western Australia, where the state was boasting abundant water supplies as late as 2000, there is now intense pressure to set aside more water for environmental purposes. The primary policy mechanism by which this has been accomplished in all states is through the reassignment of priority claims to the environment and distinguishing the agreed environmental quantum of water from other consumptive uses.

The apparatus by which the environmental allocation or share has been determined varies significantly between jurisdictions. Nevertheless, a number of common features are identifiable. First, there has been an attempt to draw upon the available science to model or predict the response of ecosystems to altered water allocations. For instance, Pagan (Chapter 14) notes the use of the MDFAT simulation tool during the development of the Living Murray response. This has proven taxing for scientists insomuch as they have been forced to reduce complex and partially understood ecological processes to a level at which the community can be engaged (see Hillman, Chapter 9). It has also fostered an unfortunate predilection to frame policy in terms of the quantity of water per se, often at the expense of other considerations like the timing of seasonal releases to achieve environmental amenity. Nevertheless, this has yielded some success, but has also provided a basis for criticism from those seeking to resist any reallocation of the resource. Second, the process has been marked by, at best, partial modeling of the social consequences. As Syme and Nancarrow (Chapter 15) observe, social criteria have been employed primarily as constraints—environmental adjustments can occur but they need to be adjudged as socially palatable, even though no serious attempt has been made to define this concept with precision. Third, the process has been characterized by occasional ad hoc political decisionmaking, particularly when resolution between competing interests has proven troublesome, or when the temptation to engage in political opportunism has proven too great. Finally, there has been an emphasis on estimating the economic impacts of reassigning water to achieve environmental amenity. By and large, much of this modeling has focused on the costs attendant on irrigated agriculture and the mechanism by which to achieve a reallocation at least cost (see Rolfe, Chapter 13).

This strand of inquiry into low-cost adjustment mechanisms has overlapped with the second main stream of policy development—the thrust toward achieving greater efficiency in the use and allocation of the resource within the consumptive pool. In this regard, pricing, water markets, and water rights have dominated the reform debate. Thus, while it is possible to conceptualize the enhanced water rights on offer to irrigators in most jurisdictions as the political trade-off for establishing an environmental reserve, it is also consistent with the desire to garner more efficient outcomes in the productive use of the resource. Moreover, following the approach described in the National Water Initiative (NWI) should result in expanded trade (see McKay, Chapter 4; Quiggin, Chapter 5) and a closer alignment of the marginal benefits and costs across multiple users (see Edwards, Chapter 10).

However, the drive toward economic efficiency within the consumptive pool has not come without its problems. Defining all dimensions of water property rights has proven particularly problematic and policy gaps have often only been revealed after the market has been given sanction by government. There is also a contentious view that efficiencies are being used as a guise to bestow state resources on particular groups at the expense of others. This is, in part, fueled by the inconsistent treatment of urban users, who are increasingly subject to state control, and irrigators, who now have stronger rights over the largest portion of the consumptive pool (see Edwards, Chapter 10; Crase and O'Keefe, Chapter 11). Of course, economists have long argued that the original assignment of rights is immaterial insomuch as trade will ultimately resolve differential valuations of the resource. In practice, the original assignment of rights to facilitate trade is always likely to be of intense interest, particularly for those with most to lose from any modification to the status quo.

This should not imply that irrigators have accepted the new policy regime willingly and that the drive toward efficiency within the consumptive pool stands to benefit this group in the longer term. As Jenny McLeod and George Warne observe by drawing on their own experiences with Murray Irrigation, major challenges remain for Australia's irrigators. First, the pace of reform has been frenetic and many within this sector claim to be besieged by constant legislative and administrative adjustments. Second, there is genuine concern that the irrigation sector is portrayed as the environmental villain in the current debate. The historical review offered by Warren Musgrave (Chapter 3) and the legal dimensions detailed by Jennifer McKay (Chapter 4) attest that the very existence of the irrigation sector is itself a manifestation of earlier policy decisions taken in good faith. Third, the irrigation sector is arguably being forced to abandon the original intentions that underpinned its genesis. For example, consider the dilemma of dealing with stranded assets. On one hand, attempts by irrigation entities to limit the permanent trade of water out of irrigation schemes can be conceptualized as a purposeful effort to restrain the full benefits of trade. Moreover, the NWI specifically tackles this problem by insisting that such restraints be progressively lifted. However, these same rules were once accepted by governments as a reasonable mechanism for dealing with the wider social objectives assumed by private irrigation companies, like Murray Irrigation, following the demise of government ownership. Incorporating a historical perspective places the reaction of irrigation companies in a different light.

Similarly, the current status of the hydroelectricity sector can be scrutinized with reference to its earlier performance and goals. In this regard, Ronlyn Duncan and Aynsley Kellow (Chapter 8) note that the hydroelectricity sector has been subject to competitive reform on several fronts. In addition to the shift toward competition in the water sector, the hydroelectricity utilities now find themselves in a competitive output market. In line with the policy thrust toward efficiency within the consumptive pool of water resources, generators have been increasingly obliged to adopt a market orientation to their activities. Duncan and Kellow (Chapter 8) remind us that this need not always be consistent with the environmental goals sought by policymakers.

With regard to the environmental objectives of Australian water policy, two broad questions remain unanswered: Has sufficient water been allocated to achieve the requisite environmental changes? And, if not, how can adjustments be incorporated into future policy decisions?

It is not feasible to provide a definitive response to the first of these questions. The variability of the resource itself and the extent of the allocation to consumptive pursuits vary markedly between and within jurisdictions (see Fletcher and Powell, Chapter 2). Although all jurisdictions have invoked planning processes to ensure an adequate allocation is provided to the environment, the urgency of these measures is not homogenous across all areas. In cases where the resource has been seriously overallocated there is little doubt that the current provisions will be found wanting. The fact that the *Living Murray* assigned 500 gigaliters and described this as the first step provides at least some support for this view. The upshot is that more water is likely to be demanded for environmental purposes in the future, particularly in the most developed valleys and aquifers.

The second of these questions is partially answered by the assignment of risks under the NWI. However, caution needs to be exercised in this regard. As several authors have noted earlier (see, for example, Quiggin, Chapter 5) the practical implementation of the risk assignment framework has yet to be tested. McKay (Chapter 4) also notes the potential difficulties of this approach from a legal perspective. It seems likely that Australia has not yet fully developed the instruments for dealing with all elements of the environmental aspects of water policy.

Technology has also been presented as offering some solutions to the need for future allocations to the environment. However, the impact of these measures is primarily at the margin when considered against the volume of extractions from agriculture (see Crase and O'Keefe, Chapter 11).

Progress against the economic efficiency goals is similarly mixed. Notwithstanding ongoing reservations about the adequacy with which rights have been defined and distributed, trade and market mechanisms are now consistently regarded as the most acceptable vehicles for achieving efficiencies within the consumptive pool. Some constraints remain, particularly between irrigation and urban sectors, but it is difficult to see these withstanding the rising demands for water access in cities.

Tensions also exist between the irrigation and hydroelectricity sectors (Duncan and Kellow, Chapter 8). Here discord arises largely because the rights to water are defined primarily in volumetric terms and hydroelectricity has hitherto been broadly regarded as a non-consumptive user. Because the economic returns to hydroelectricity are not always optimized by dam releases that suit other extractive (and nonextractive) users, and because hydroelectricity has been granted priority rights, there is no need for generators to consider the demands of others. Some trades to secure changes in the timing of releases from hydroelectricity storages have been observed, although a wider market for such rights remains undeveloped.

Active interest in technology can be found, particularly in the urban sector where adjuncts to potable water supplies are being sought (see Khan, Chapter 12). Reuse of urban wastewater is high on the agenda, although the absence of markets for reclaimed water will be a challenge. Arguably, better definition of the rights to amenity in the environment that receives a wastewater stream might assist, insomuch as this would produce different price signals to capture the merits of the reuse technology. Changes to the way in which urban water pricing is conceptualized may also be required. However, this approach offers greatest appeal in Sydney, because the prospects for intersector trade are most limited in this case. In other cities, the economic benefits of intersectoral trade are proving increasingly difficult to refute, either on social or environmental grounds (see Edwards, Chapter 10).

On the social front, there has been virtually no attempt to seriously incorporate social or cultural values into policy goals. Rather, the aim would appear to have been restricted to encapsulating fairness into the processes that inform the allocation of water to the environmental pool. Arguably, this limited approach will make it increasingly difficult to promote alternative technologies in the urban setting (see Khan, Chapter 12), although this may be offset by the penchant for political intervention in urban water policy (see Crase and O'Keefe, Chapter 11).

What Lessons Might Be Derived from the Reform Experience?

Water policy provides insights into the difficulties of adjusting policy settings generally and offers potential lessons for reform both in Australia and elsewhere. As we have seen, the progress in the last two decades has been substantial in response to both environmental and economic challenges. States and territories have independently undertaken actions to secure a separate pool to meet environmental ends. They have also managed to overcome jurisdictional differences and rivalries to meet collective objectives in this context, albeit assisted by national suasion and financial incentives. This is not a unique policy accomplishment although there are many other nations who have yet to confront this dimension of water policy seriously.

The urgency to deal with environmental claims on Australian water is a function of two underlying factors. First, the inherent variability of the resource has meant that human actions to produce a secure water supply have had a profound impact on natural ecosystems. Flora and fauna well-suited to variability have been threatened and, in some cases, driven to extinction. Human action also threatens the quality of the resource. Second, the institutional arrangements that drove the developmentalist approach attracted strong public support in all states for at least sixty years. The combination of these two phenomena has only recently manifested in greater attention to environmental water allocations. Moreover, this policy approach might be less likely to arise where these two forces hold less sway.

Nevertheless, countries with less variable hydrology and/or modest consumptive behavior can still learn from these experiences and should guard against complacency. In particular, the need to develop robust adaptive arrangements for the allocation of the resource cannot be understated. After all, the values ascribed to the environment seem likely to change in all societies over time—observed responses to changing income, demography, and technology support this view.

In contrast to the need for adaptive institutions to deal with expanded environmental demands, there is also a clear need for the capacity to cater for changes on the efficiency front. In this context, Australian policymakers have opted for property right regimes that are amenable to a market setting, at least in the case of the most profligate users. *Prima facie*, this is at odds with the desire for adaptive arrangements in the management of environmental reserves. However, as has been shown in Australia, this need not be the case. Careful crafting of policy and adequate consideration of all dimensions of property rights can give rise to adaptive institutions capable of promoting efficient use—these goals should not be considered mutually exclusive.

Another important lesson from the move toward a market framework relates to the sequencing of policy adjustments. Few economists would dispute the potential efficiency enhancements that attend the market when rights are adequately defined. However, one of the recurring themes of Australian water policy has been the struggle to develop an adequate definition of rights capable of encompassing all dimensions of water use. In some cases this can be traced to the initial CoAG reforms where strong financial incentives were offered to each state for the establishment of a water market. Whereas the early reforms sought to break the nexus between water and land and to foster the establishment of a market in water entitlements, closer attention to the precise definition of rights and essential water accounting did not appear on the policy landscape until much later. Little wonder that policymakers are only now starting to grapple with thorny issues like the impacts of trade on third parties.

In policymaking utopia the decisionmaker would have sufficient information to fully understand the relationship between each poorly defined element of a water right and any resulting welfare losses. In addition, the omniscient policymaker would be blessed with the necessary foresight to systematically compare potential welfare gains against the costs of designing institutions that adequately enhance those rights. This has not been the case in Australia, nor is it likely to reflect the true policymaking environment elsewhere. Nevertheless, some general conclusions can be drawn from the Australian experience to date.

First, the broad substantiation of volumetric rights for irrigators within state jurisdictions has yielded substantial gains once trade has been permitted. The sheer volume of seasonal trade in agriculture stands as testament to the enhanced allocation outcomes on this front and, as noted by McLeod and Warne (Chapter 7) and Rolfe (Chapter 13), trade has become a critical component of farmers risk management strategies. The gains to the urban sector have not been fully realized and are unlikely to be so in the immediate term. Permanent water trade, although initially stilted, shows signs of expanding, although arguably its modest growth attests to the fact that this is where many of the property rights challenges reside.

Second, the task of articulating the impacts of trade on supply reliability looms as a major challenge, particularly if the sequence of low rainfall years being experienced over the last decade in southern Australia persists. In this context, and with the benefits of hindsight, it may have been more propitious to devote greater policy effort to restoring water extractions to more sustainable levels before promoting the extent of trade that presently exists. Strongest support for this argument is evident in New South Wales where efforts to reduce water extractions are proving costly on many fronts. The institutional coupling of a broad water market with the implementation of sustainability thresholds to achieve environmental ends has not yet been fully tested. The current thrust toward a national water market accompanied by the longest sequence of dry years on record will likely prove to be a rough ride for many. A scenario of continued drought accompanied by downstream water purchasers attempting to invoke water rights which later prove impossible to deliver is not a prospect that supporters of water markets would relish. Hopefully, the enthusiasm for a market framework at the federal and state levels has not progressed so far ahead of the scoping of rights as to ultimately undermine its usefulness.

Third, considerable national attention has been placed on the rights of the irrigation community to retain a viable irrigation network. The extensive policy activity that circumscribes the appeasement of these interests belies its significance. As noted earlier (Crase and Dollery, Chapter 6, and Edwards, Chapter 10), there is a strong case for arguing that such rights are illusory and adequate policy measures are already in place to attend to these concerns.

Fourth, the necessity for measures to deal with environmental externalities arising from trade has mostly been revealed after the event. This is as much a reflection of the incomplete state of knowledge about Australia's riverine and groundwater ecosystems as it is symptomatic of a lack of foresight on the part of policymakers. By and large, most of these problems have proven tractable and, as additional information has been brought to light, appropriate responses have been put in place. Arguably, the retention of robust superordinate rights in the state has proven useful in this context and the costs of forestalling trade would have overwhelmingly exceeded any benefits of waiting for more information.

A final observation relates to the greater and widespread economic prosperity enjoyed by Australians throughout much of this final phase of water policy. Notwithstanding that fiscal stringencies originally encouraged states to sever, or at least distance themselves from irrigation (and in some cases urban water) infrastructure, the unprecedented economic well-being of Australians has afforded a relatively unique environment. In some instances, public funds have been available to act as a lubricant to reform (although the efficacy of some of this spending has been called into question in some earlier chapters). Prosperity also provides the luxury of considering the long-term status of the environment—an option not always readily available in more populous or poor countries. Arguably, impetus has also been added by the sequence of low rainfall years over the past decade, which has accentuated the environmental challenges and simultaneously signaled the relative economic disadvantage caused by tying resources to irrigated agriculture.

Regardless of exogenous events, paramount to the refinement of water policy is the recognition of path dependencies and their role in the future. In essence, today's policy choice becomes tomorrow's policy constraint, particularly if property rights have been carelessly assigned in a manner that ignores the finite nature of the resource and likely future demands. Nations presently on a path toward increased water resource development should take particular heed from the Australian experience. Assigning the rights to water during the expansion phase can impose significant costs in a mature water economy. There is also no guarantee that the economic fortune that has circumscribed the recent phases of Australia's water history will accompany the necessity for adjustment in other countries.

There is a long-standing maxim attributed to Mark Twain which posits that "whiskey is for drinking and water is worth fighting for." Careful attention to the assignment of rights can at least mitigate some of the conflict that potentially attends competition for this vital resource. The experience in Australia shows that this is no easy task but, nonetheless, one which offers significant payoffs.

Notes

1. Notable exceptions to this occur in the areas of national defense and responses to terrorist threats.

Index

T - #0123 - 101024 - C0 - 234/156/15 [17] - CB - 9781933115580 - Gloss Lamination